KU-300-965

WITHDRAWN

STIRLING UNIVERSITY LIBRARY
3 0042 00326 1977

Soiling and Cleaning of Building Façades

Soiling and Cleaning of Building Façades

Report of Technical Committee 62 SCF
RILEM
(The International Union of Testing and Research Laboratories
for Materials and Structures)

Edited by

L.G.W. Verhoef

London New York
CHAPMAN AND HALL

690.24 VE

First published in 1988 by
Chapman and Hall Ltd
11 New Fetter Lane, London EC4P 4EE
Published in the USA by
Chapman and Hall
29 West 35th Street, New York, NY 10001

@ 1988 RILEM

Printed in Great Britain at the
University Press, Cambridge

ISBN 0 412 30670 0

All rights reserved. No part of this book may
be reprinted, or reproduced or utilized in any
form or by any electronic, mechanical or other
means, now known or hereafter invented,
including photocopying and recording, or in
any information storage and retrieval system,
without permission in writing from the
publisher.

British Library Cataloguing in Publication Data

Soiling and cleaning of building façades.
1. Buildings. Exteriors. Cleaning
I. Verhoef, L.G.W. (Leo G.W.)
690'.24

ISBN 0-412-30670-0

Library of Congress Cataloging in Publication Data

International Union of Testing and Research
Laboratories for Materials and Structures.
Technical Committee 62 SCF.
Soiling and cleaning of building facades :
report of the Technical Committee 62 SCF,
RILEM (the International Union of Testing
and Research Laboratories for Materials
and Structures) / edited by L.G.W. Verhoef.
p.   cm.
Bibliography: p.
Includes index.
ISBN 0-412-30670-0
1. Exterior walls–Cleaning.   2. Weathering of
buildings.   3. Building materials–
Cleaning.   I. Verhoef, L.G.W.   II. Title.
TH2235.I58 1988
648'.5–dc19   88-11823

Contents

Members of Technical Committee 62   vii

Acknowledgements   ix

Introduction   xi

1 The skin of the façade   1
   1.1 The skin of concrete   2
   1.2 The skin of stones and bricks   8
   1.3 The skin of wood   10
   1.4 The skin of glass   12
   1.5 The skin of plastic building components   14
   1.6 The skin of metals   15

2 Changes in the appearance of the skin of the façade   18
   2.1 Introduction   18
   2.2 Changes caused by atmospheric constituents   18
   2.3 Changes of aspects of façades due to particular substances present in building materials   49
   2.4 Façade alterations caused by adjacent and additional materials   51
   2.5 Changes by soiling   58

3 Non-biological soiling   65
   3.1 Airborne particulate matter and its measurement relevant to the soiling of façades   65
   3.2 Transference of atmospheric pollution to the façade   83
   3.3 Attachment of particles to the façade   92
   3.4 Rainfall run-off   98

4 Biological soiling   111
   4.1 Introduction   111
   4.2 Viable particles in the air   112
   4.3 Conditions for microorganism development   115
   4.4 Effects of microorganisms on substrates   117

5   Cleaning of soiled façades   124
    5.1 Introduction   124
    5.2 Some physical principles of cleaning   126
    5.3 Cleaning methods   127
    5.4 Problems involved in the treatment of soiling of biological
        origin   132

6   Recommendations for design of building façades   136
    6.1 Introduction   136
    6.2 The situation (location)   136
    6.3 Façade material   138
    6.4 The profiling of the form and the detailing of buildings   140

Appendix 1   The sandblast test as a method to judge the properties
of the surface   155

Appendix 2   Test results of the 'soft' sandblast method   157

Appendix 3   A method for determining the soiling capacity of air   162

Appendix 4   Test results of the reflectance measurements   165

Appendix 5   Test methods on biological soiling   172

Appendix 6   Outdoor exposure tests of atmospheric soiling   174

Appendix 7   Water absorption test   178

Index   185

Members of Technical Committee 62 Soiling and Cleaning of Building Façades

The late Professor E.M. Theissing
(Chairman)

F. Hawes Dip Arch RIBA
Senior Advisory Architect
Cement and Concrete Association
Slough
United Kingdom

J.M. Estoup
Ingénieur CNAM
Centre d'Etudes et de Recherches de l'industrie du Béton Manufacturé
(CERIB)
Epernon
France

A. Perrichet
Docteur Ingénieur
Laboratoire de Géoméchanique Thermique et Matériaux
Institut National des Sciences Appliquées
Rennes Cédex
France

Ulrich Trüb
Arbeitsleiter Technisch Forschungsund Beratungsstelle der Schweiz Cement-
Industrie
Wildegg
Switzerland

K.L. Pwa
(Secretary until September 1985)
Senior Lecturer
Group Materials and Constructions
Department of Civil Engineering
Technical University Delft
The Netherlands

C.A. Kortland
(Secretary from September 1985)
Senior Research Engineer
Central Laboratory Bredero
Maarssen
The Netherlands

L.G.W. Verhoef
Senior Lecturer in Renovation and Maintenance Techniques
Department of Architecture
Technical University Delft
The Netherlands

Acknowledgements

Contributions have also been received from specialists outside the committee and their assistance is gratefully acknowledged:

O. Beyer Msc., Project Leader of the Cement-Och-Betong-Instituet, Stockholm, provided the sections on rainfall run-off.

Dr R.D. Degeimbre, Senior Lecturer of the Liège University in Belgium, provided the sections on the skin and weathering of plastic.

Dr P. Rossi-Doria, CNR Centro 'Conservazione Opere d'Arte', Italy, provided the sections on the skin and weathering of stones and bricks.

Dr J.F. van de Vaate, Scientific Director of the Dutch Energy Research Foundation, provided the section on airborne particulate matter and its measurements relevant to the soiling of façades.

M.W. Verver, of the Materials and Constructions Group, Delft University of Technology, provided the sections on the skin and weathering of wood.

Professor H. de Waal and N. van Santen, of the Institute of Applied Physics TNO-TH glass department provided the sections on the skin and weathering of glass.

Professor J.H.W. de Wit, Delft University of Technology, provided the sections on the skin and weathering of metals.

Mrs Willy Harkema-Louw and Mrs Martha Betten-Venhuizen of Bouwcentrum Technology, Maarssen, prepared the final typescript.

The help of Dr Sjef Schoorl from the applied linguistics department, Delft University of Technology, is acknowledged for his valiant attempts at bridging the larger gaps between the language used in the manuscript and standard English. Thanks are also due to Mr Michael Dunn, Ms Madeleine Metcalfe and the editorial and production staff of Chapman and Hall for their assistance.

Introduction

Rilem Technical Committee 62 SCF was formed under the chairmanship of the late Professor Eric Theissing to study the soiling and cleaning of building façades. In order to make useful recommendations for suitable methods of cleaning, it was necessary to investigate the complex system by which surfaces are changed by the natural and unnatural elements in the environment. Some of these changes are physical and others superficial but all need to be understood. It is necessary to study the ways in which pollution in the atmosphere is transported to buildings and adheres to surfaces, and the manner in which the dirt can subsequently be redistributed by water on the façade. This redistribution of the dirt is normally responsible for the most noticeable visual changes, so the movement of wind around buildings and its influence on the quantities of rain which hit different parts of the various façades also forms part of the study.

Dirt affects façades in a number of ways. Some types of soiling can physically damage certain types of surface but in many cases there is only a visual change. Some materials, because of their colour or texture, can look acceptable in spite of being dirty while others only look right when they are clean. However, visual changes are sometimes so great as to have an adverse effect on the value of a building. These phenomena are therefore worthy of study so that they can be anticipated by the architect and taken into account at the time that the building is designed.

Buildings can be designed to minimize dirtying, but the study of rain and wind movements around buildings shows that in most cases it is impossible to ensure that all parts of all façades receive enough rainwater to keep them clean unless the building can be designed in a form resembling a pyramid. It would be quite absurd to suggest that all large buildings should be pyramid-shaped so we are forced to accept that some parts of such buildings will not be naturally cleaned.

The architect may decide, by choice of materials and by providing access, to make these parts of the building easy to clean, but there is another option. This is to accept that these areas will remain dirty and, by careful choice of materials and details, the architect can attempt to ensure that the building can carry this dirt without being physically or visually harmed. Many of the old buildings in our towns and cities are in this state of being partly clean and partly dirty and are worthy of serious study to determine to what extent the soiling is spoiling their appearance (see e.g. St Paul's Cathedral).

By investigating the environment in which a building is to be built – its microclimate and its pollution – and then bearing these conditions in mind in the design of the building, it is possible for an architect to predict those parts of the building which can be kept clean and those which will remain dirty. Materials can then be chosen which are suitable for these situations and which will control the flow of water on the façade, so that there will be no dirty streaks on the clean surfaces and so that the dirt in the other areas will remain undisturbed.

All of these subjects are touched upon in this report together with the most suitable methods for cleaning the various building materials mentioned. Finally, a number of appendices describe experimental test methods related to the study and measurement of the soiling of various surfaces.

St Paul's Cathedral

Chapter One

The skin of the façade

When a building material is in contact with contaminants, the nature of the surface of the material is of great importance for the adhesion and conspicuousness of the contaminants. Moreover the capillar system of the outer layer has important implications for water absorption and for the capillary condensation of water vapour.

The properties exhibited at the surface of a building material at the production or delivery stage may be quite different from the average mechanical, physical and chemical properties of that material as they are known to the architect or civil engineer.

The layer of deviant composition at the surface is often described as a skin. The layer is sometimes very thin. For stainless steel it amounts only to about 10^{-2} μm. For concrete the thickness of the skin is somewhat arbitrary as the changes in composition of the interior concrete take place gradually, but it certainly amounts to 1 mm or more.

For many concrete elements used in the façade this layer will have been removed at the production stage.

Consequently, directly after a concrete slab with exposed aggregate has been produced, no skin may exist.

However, after a period of weathering a new skin may be formed as a result of carbonation, by the action of rainwater in selective solution on the components of the cementstone and by the action of salts or gases in the air.

Because of weathering the skin of different building materials will also react differently or even contrariwise. Thus the skin of concrete disappears over the years and the true concrete becomes visible while real fresh natural stone is obscured and a natural stone skin forms.

The production method and the process of weathering thus determine the skin of a building material.

The properties of a building material are of ultimate importance for the adherence of paints, glues, seals and dirt and also for the resistance to wear of the material.

Moreover, for a proper choice of cleaning method to be made, knowledge of the nature and chemical resistance of the skin is indispensable.

In this chapter a description will be given of the skin of some building materials used in façades.

Façade materials can be broadly categorized according to their different values of porosity, and can thus be divided into :

    (a) porous materials : concrete ; bricks ; natural
                      stone ; wood
    (b) non-porous materials : glass ; plastics ; metals

Since paints and seals are renewed several times in the lifetime of a façade they will not be mentioned here.

For many building materials there is in general a lack of well-standarized methods for the determination of the properties of the façade. Furthermore, the number of investigations is limited. Thus it is not always possible to give clear information on the outer layer as a function of production and weathering.

Where possible, values should be obtained for : the thickness of the skin ; the light reflection of its surface ; the hardness resistance to wear ; the porosity of the outer layer or roughness of the surface ; the water suction and water permeability ; the gas diffusivity ; the chemical and mineralogical composition.

1.1 THE SKIN OF CONCRETE

1.1.1 Description of the skin

After concrete has been manufactured its composition, considered over the cross-section of the constructional element, is not uniform. This applies over the horizontal as well as over the vertical sections. One important cause of this phenomenon is the segregation due to gravity and the compaction process. If this process is obstructed by the reinforcement or by the mould, the heterogeneity will be strengthened locally.

The other cause of heterogeneity originates from the wall effect. On the subject of the bleeding process and the wall effect there is a wide literature. It is still surprising, however, that so few data are available, from the existing gradients in actual concrete elements, of the water-cement ratio, the aggregate-cement ratio and the gradation of the aggregates, described as a function of the distance from the surfaces of the element.

Kreyger (1) has investigated 2 mm slices sawn from bottom and lateral planes of concrete specimens of 100x100x500 mm.

If the W/C ratio and A/C ratio are known it is possible to estimate the concrete properties as a function of the distance from the outer plane (Figure 1.1.1).

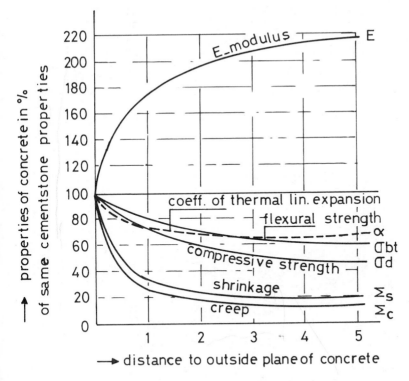

Figure 1.1.1.

The concrete contained 325 kg cement/m^3 ; the
W/C =0.54 and the A/C = 6.11.
In the outer 2 mm layer at the lateral planes, the A/C
ratio was found to be 3.1. The water absorption (in 10
min) for this slice of concrete amounted to 20,0 vol.%
whereas the A/C of the slide from 3-5 mm was 5.55 and the
water absorption was 11 vol.%.
It seems that under these circumstances the difference
in concrete at the perimeter at a depth of 6 mm, and that
at the core was only very small. In the first 2 mm,
however, a sharp gradient in the properties is to be
expected.
As most of the weathering of normal concrete and also
the soiling relate only to a depth of not more than 1 mm,
a proposal can be made to restrict the real area of the
skin of concrete to 1 mm.
Methods supplying information on the gradients in this
layer must be sought out.
One attempt was made by Deelman and Kreyger (2) to
define a surface by determination of the microroughness.
The instrument (talysurf 4) used records the movements in
the vertical direction of a diamond needle travelling
along horizontal lines. The dimensions of the truncated

3

top of the pyramidal needle are 1.2 x 1.2 μm. The surface
of the concrete for analysis must be polished. The A/C
was determined in a plane perpendicular to the surface.
In Fig. 1.2 the results show that at a distance of 0.1 mm
from the surface there was still about 20 vol.% present.

At a distance of 1 mm the amount of aggregate was 60
vol.% and at 4 mm and greater the quantity of aggregate
had reached its bulk value. The results obtained for the
A/C ratio were roughly in agreement with optical
measurements.

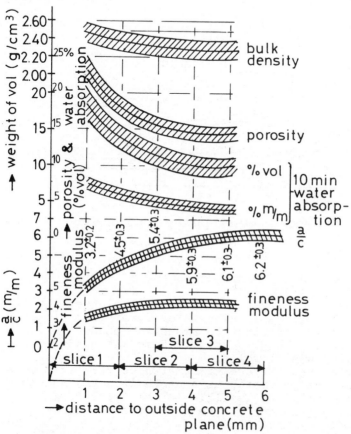

Figure 1.1.2.

A knowledge of the A/C ratio and of the W/C ratio are
necessary for prediction of the shrinkage of the skin.

Haircracks are often visible on the surface of
concrete. The depth is mostly less than 1 mm, so it has
to do with the properties of the skin. If dust is
deposited in its starform pattern it can spoil the
appearance.

If the concrete slab is dark in colour then the

4

haircracks can become visible due to lime effloresence.
Determination of the original W/C ratio of the skin is
also possible in principle by measuring the microroughness, as the wider pores of the hardened cement paste
are greater than 1.2 μm.
Another method introduced by the author (see Appendix
1) is to determine the hardness of the concrete by
sandblasting the surface at a certain pressure so that
only a layer of, for instance, 50 or 100 μm is stripped
after several minutes of blasting. Thus the hardness of
the skin is determined over very small distances from the
surface. If the A/C ratio is also known in this layer it
is possible to compare the measured hardness with that of
cement pastes of the same A/C ratio (for instance 20% fine
sand) but different W/C ratios cured in the same way as
the original concrete. When curing took place at 95%
R.H., the W/C ratio of the skin of a concrete cube with a
W/C ratio of 0.5, was more than 0.6.
Using this method it could also be established
quantitatively that a dark-coloured area on a given
concrete slab is much harder, due to a lower W/C ratio,
than a lighter coloured area (see Appendix 2). These
differently coloured areas often occur on prefabricated
slabs when the stiffness of the moulds is insufficient or
the vibration is not uniform.
The phenomenon of the presence of dark areas on
concrete is often described as clouds. Cement hydration
is slower when the W/C ratio is low. Unhydrated cement is
dark in comparison with hydrated cement. Also, the total
cement content at the surface is greater when the W/C
ratio is low. In addition to the fact that the dark area
was much harder than the light-coloured area it appeared
that there was a marked difference in the sand gradation
at distances up to 3 mm beneath the surface :

Distance from surface	Maximum size of sand/gravel (cross section)	
	Dark location	Light location
50 μm	0.4 mm	1.5 mm
150 μm	2.0 mm	6.0 mm
300 μm	4.0 mm	8.0 mm
900 μm	5.0 mm	10.0 mm
2700 μm	10.0 mm	12.0 mm

As these two areas were situated next to each other the
conclusion is reached that gradients in W/C and A/C ratios
also somtimes exist parallel to the surface and not only
perpendicular to the surface. Sometimes air voids could
be detected underneath the closed dark surface from 50 μm
to 90 μm.
From the compositions determined it is possible to
calculate or to determine the properties as strength,

5

creep, E-modulus, shrinkage, water vapour diffusivity, water absorption and capillarity.

The water vapour diffusivity has also been determined directly by the author for thin slices only 1.5 mm thick taken from a concrete specimen measuring 100 x 400 x 300 mm^3.

It is interesting to see that the outer layers up to a depth of 6 mm are less permeable than the core and that of the 3 mm the first mm is the most permeable.

This fact is shown in the following table :

Water vapour diffusion resistance factor

From - To	Horizontal plane of an element μ	Vertical plane of an element μ
0.0 mm - 1.5 mm		24
0.5 mm - 2 mm	69	51
1.0 mm - 2.5 mm	83	68
1.5 mm - 3 mm	75	x
2 mm - 3.5 mm	59	x
5 mm - 6.5 mm	48	52
20 mm - 21.5 mm	45	x
100 mm -101.5 mm	26	x

In conclusion, it may be said that at the surface of concrete after a curing period of 28 days a layer of ± 1 mm with less than 20% fine aggregate exists. The W/C ratio of this layer depends on the conditions during production. In light coloured areas the outer 150 μm of this mm has a somewhat higher W/C ratio, as in the core. The layer is easily sandblasted away. In darker coloured areas the top layer has a greater hardness and a lower W/C ratio.

The layer at a distance from the surface of 1 mm to 2.5 mm has the lowest vapour permeability. This property increases in value at the upper surface and also towards the concrete core where larger particles of aggregate are also present.

It is not possible to generalize here for all types of concrete. An investigation is needed to establish how the structure of the skin is determined by the mix design and conditions during manufacture.

References

(1) Kreyger P.C., (1984) The skin of concrete - Composition and properties. <u>Materials</u> <u>and</u> <u>Structures</u>, 100 275-283.
(2) Deelman J.C., (1984) Textural analysis of concrete by means of surface roughness measurements. <u>Materials</u> <u>and</u> <u>Structures</u>, 101 359-367.

1.2 THE SKIN OF STONES AND BRICKS

Quarry stones and bricks after manufacturing do not show great differences in the basic characteristics (such as porous structure, chemical position, mechanical resistance, etc.) when superficial layers and inner parts are under consideration. Ancient bricks and most of the modern ones are very homogeneous ; only some particular kinds of modern bricks show an outer skin of few millimeters, highly vitrified and hardened.

Differences in the characteristics of outer and inner parts arise when the materials are used as components in a building. Atmospheric agents and pollution act on the exposed surfaces and a superficial layer is generally produced with chemical, structural and mechanical characteristics different from those of the internal part.

The external surface can also show macroscopic alteration, such as chromatic variations, cracking, fissuring, pitting, chipping and so on.

Crusts and efflorescences can be present as a result of the pollution and of the migration of soluble salts.

The skin of stones and bricks as a layer of a certain order of magnitude is not easily defined ; the thickness of this layer can be very different according to the mechanisms of alteration, the chemical composition of the materials and, mostly, to their structural characteristics (1). Stones show an extreme variability in porosity, pore-size distribution and specific surface : the literature gives values of the total porosity of 0.5-42% for sandstones, 0.8-27% for hard limestones, 0.1-6% for marbles. The porous structure of bricks can also be very variable according to the influence of many different factors such as the chemical and mineralogical composition of the raw materials, their ratio, the firing conditions and so on.

Repair and restorative interventions can be applied, from the external surface of the materials, on layers of different thickness according to the aim of the intervention itself and to the methods used.

Cleaning, which is intended for removing whatever is damaging the stone and the brick, should concern only the outer layer composed mainly by non-stony materials, such as incrustations, soot particles, microorganisms and so on.

Surface treatments by chemicals are applied in order to retard the deterioration processes by creating an external sacrificial layer ; they concern only a few millimeters of the skin of the original stone and/or brick.

A consolidating material has to penetrate as deeply as possible in order to improve the adhesion of altered external layers to the sound substratum, nevertheless, in the most favourable cases, the depth of penetration of a consolidant is a few centimeters.

References

(1) Rossi-Doria P., (1985) Principles and applications of Pore Structural Characterization. Bristol p. 441.
(2) Torraca G., (1982) Porous Building materials ICCROM Rome.
(3) Alessandrini G. (1981) The conservation of stone, Int, Symp., Centro "Cesare Gnudi", Bologna, Part A, p. 139.

1.3. THE SKIN OF WOOD

A general description of untreated wooden skins is much
more difficult than in the case of other types of skin.
Wood is a natural product of biological origin, and the
complex and extremely directional structure of wood tissue
makes the product's structural and textural
characteristics radically different - even to the naked
eye - from those of man-made materials such as plastics or
metals. Needless to say, it is precisely those
characteristics that determine the effect of external
influences on the wooden skin.
     Due to the complex internal cohesion of wood
tissue, the appearance and behaviour of a given piece of
wood will be determined by a variety of factors, such as
the species of tree from which the piece was cut (softwood
versus hardwood), its gross anatomical features (sapwood
or heartwood, wood rays, growth rings, resin ducts, and
the characteristic anisotropy of wood), its plane of view,
the way in which it was processed (sawing, planing
splitting), and its age.
     The major distinction between different species of
tree is between softwood and hardwood trees. Softwoods and
hardwoods have different structures, which result in
different surface properties. Hardwoods are characterized
by the presence of vessels, whose sole function is to
conduct water through the tree. Softwoods lack vessels.
Instead, they have tracheids: elongate cells that combine
water-conduction functions with supporting functions. the
anatomy of softwoods, therefore, is less complex than that
of hardwoods.
     Sapwood (the light-coloured wood from the outer layers
of a tree stem) and heartwood (the darker wood from the
inner part of the stem) may react differently to external
influences. Heartwood offers greater resistance to decay
than does sapwood, due to the presence of certain toxic
substances in heartwood. A complicating factor here is
that the proportion of sapwood to heartwood varies from
species to species and from tree to tree.
     Growth rings (the layers of wood that grow in the
course of one season) contain two types of wood, with
different characteristics: light-couloured springwood,
formed in the early part of the growing season; and darker
summerwood, formed towards the end of the season.
Springwood has a higher proportion of large, open thin-
walled (ca. 2 μm) cells, and is accordingly more porous.
     Summerwood has cells with thicker walls (up to 10 μm),
which makes it harder and less poro, and therefore show
less contrast between the springwood and the summerwood in
their growth rings.
     Wood rays are composed of cells that are used for
storage of food, and are found in all types of wood. In

10

hardwoods, however, these cells have thicker walls than in softwoods, where they are normally invisible. Wood rays may influence the behaviour of the wooden skins.

Because of the strong directional structure of wood tissue and the resulting anisotropy of wood, the study of wood anatomy focusses on three different planes of view, a cross-sectional plane, and two longitudinal ones : a radial plane of view, obtained by making a longitudinal section at right angles to the growth rings, and a tangential plane, growth rings will usually show up in a parabolic shape.

Since these different sections will bring different aspects of the wood tissue to the surface of a piece of wood, it is easy to see that different planes of view in wooden skins will have a different effect on skin behaviour.

Woodworking methods used in the production of wooden skin will also be of influence, since they will lead to surfaces that can be either rough or smooth, have little or much porous tissue in it, etc.

Untreated wooden skins will change in the course of time mainly because of chemical changes in the materials (cellulose and lignin) from which wood is built, but also because of shrinkage cracks, which may occur in different directions, and which will increase the total surface of the skin exposed to external influences. In addition, the strong hygroscopic character of wood may occasionally result in drastic changes in the behaviour of wooden skins.

1.4 THE SKIN OF GLASS

The skin of glass, defined as a surface layer of the order of 0.1 - 1 μm, has a chemical composition different from that of the bulk. Reactions such as volatilization during formation, thermal treatment, weathering and cleaning may reduce the concentration of such elements as alkalis, resulting in a silica - rich surface.

The unsatisfied Si-O- and Si- bonds have a great affinity for water, forming a hydrated surface layer of SiOH groups.

Relatively thick leached surface layers result from corrosion, typified by a hazy appearance and affecting the light transmittance, reflectance and strength. Glass corrosion will be discussed in chapter 2.2.5.

The mechanical condition of the glass surface, especially the presence of small cracks or flaws spread out over the surface of glass products, has a great effect on the strength of glass. Cracks and flaws are introduced by production technology, heat treatment, polishing, handling etc., and are normally invisible to the eye.

The strength degradation of brittle materials like glass depends on the size of the flaw. When stress is applied the stress at the crack tip increases by a factor $2\sqrt{d/\rho}$ where d is the crack depth and ρ is the radius of the crack tip (1,2).

The strength degradation of glass also depends on the water vapour concentration in the atmosphere because of stress-induced corrosion due to water present at cracks tips (3).

Because of the mechanical defects at the surface of glass, the real strength of glass is less then 1/200 of the theoretical value of strength of 16 000 MPa, and when severely abraded even lower.

Flaw size and distribution are a matter of probability, resulting in a certain variance in breaking strength.

By special treatment - tempering or chemical treatment - compressive stresses are introduced at the surface, thus reducing the effect of cracks and flaws. Strengthening by a factor of 3 - 5 can be achieved (4).

References

(1) Griffith, A.A. (1921) Phil. Trans.Roy. Soc., 221 A, 163.
(2) Wiederhorn, S.M. and Bolz, L.H. (1967) Stress corrosion and static fatigue of glass. J. Am. Ceram. Soc., 50 (8) 543-548.
(3) Wiederhorn, S.M. (1967) Influence of water vapor on crack propagation in soda-lime glass. J. Am. Ceram. Soc., 50 (8) 407-414.

(4) Littleton, J.T. (1936) Effect of heat treatment on the
    physical properties of glass. Bull. Am. Ceram. Soc.,
    15 306-311.

1.5 THE SKIN OF PLASTIC BUILDING COMPONENTS

Plastic components are generally manufactured from a formulation in the liquid state. The change from liquid to solid state occurs by cooling in the case of open chain polymers (thermoplastics) and by crosslinking reactions for crosslinked polymers and vulcanized elastomers.

The state of surface of the material will therefore depend on the state of surface of the mould or of the die when the product is moulded in its final form and aspect. For other products, there are other operations to take into account (thermomoulding, welding). For open chain polymers (thermoplastics) the steel moulds are machined and processed to obtain a surface finish which is generally smooth and dense.

For resin-based products, the mould surface may be rougher and is generally coated with a demoulding agent which may persist on the skin of the product. In certain cases, an internal demoulding agent blended in the formulation will migrate to the surface of the product.

Polymers are generally waterrepellent and the surface tension can be adapted by thermal or electric treatment.

The dielectric properties of materials are responsible for their ability to absorb static electricity and store it. This can be avoided by treatment with an internal or external anti-static agent.

1.6 THE SKIN OF METALS

The evaluation of metal surfaces cannot be carried out according to the same criteria as are used to evaluate stony materials. Moreover, classification or description will very much depend on the method of analysis.

Surface films.

A metal surface may look clean when it is observed with the unaided eye, but on close examination at microscopic level it will normally become clear that the surface is covered with a layer of composition different from that of the base metal/alloy. A relatively thick layer consisting of (hydr-) oxides, salt deposits and other contaminants is present. Moreover, the alloy composition near the surface differs from the bulk composition due to surface treatments, stress, relaxation and selective interaction with one or more ambient components. The "oxides" formed on the metal surface are not always chemically homogeneous.

For example, the stoichiometry of the oxide layer on iron/steel depends on the proximity of the steel/oxide and oxide/atmosphere interfaces, with a higher oxygen content near the outside of the layer.

At higher temperatures even as many as three separate phases may be present, $Fe/FeO/Fe_3O_4/ Fe_2O_3$/atmosphere.

Moreover, in carbon steel, depletion of the carbon in a thin surface layer of the alloy may result from fast selective oxidation of the carbon. The opposite (carbonization) may also occur in a carbon-containing atmosphere. These examples show that both the oxide layer and the top layer of the alloy may suffer from a chemically inhomogeneous composition.

Aluminium and its alloys can also be considered as passive, because they are covered with a highly resistant and protective thin oxide film (thickness 1-3 nm). The film is self-healing and it also improves the appearance of the metal. Thicker and more decorative oxide coatings can be obtained by anodizing and sealing.

On the surface of stainless steel a thin chromium (hydr-) oxide is present (1-2 nm), responsible for resistance to corrosion. The film is non-porous, insoluble and self-healing if broken under normal conditions.

Moreover, the surface of copper is covered with an inhomogeneous oxide layer, Cu_2O near the metal and CuO near air.

However, depending on the relative humidity and the purity of the ambient air, basic copper carbonates and sulphates or hydroxychlorides may also be present.

Surface porosity.

Most metal alloys are crystalline materials which contain various imperfections in their lattice structure. The absence of an atom that would be present in a perfect crystal is called a vacancy.

Vacancy condensation near the surface can result in surface porosity. Surface porosity can also exist due to the presence of a porous oxide film on the surface. The pores are then as a rule of submicroscopic size, their diameters being in the range 10-50 nm, and their direction is mostly perpendicular to the ambient oxide interface.

Surface stresses.

Stresses in surface layers differ unfavourably from those in the bulk due to several reasons.

Loading stresses produced in elastic bending are not equally distributed over the cross section and are high in the surface layer. The surface roughness which makes the surface non uniform along the main planes will cause stress concentrations of great intensity at several points along the surface.

Surface treatments may also affect the condition of the surface in two ways. Firstly, the physical properties may be changed by mechanical means, by heat treatment or by the deposition of a new surface layer, and secondly a difference between the internal stresses in the surface layer and those in the bulk may be introduced as a result of changes in the volume of the surface layer. With surface rolling, for example, the outer surface layer is under considerable compressive stresses, while these stresses are balanced by tensile stresses in the sub-surface material. Residual tensile stresses due to surface treatment, seldom the design stresses, may cause stress-corrosion cracking. (This strongly depends on the type of surface treatment : e.g. shot peening of stainless steels results in compressive stresses in the surface !)

Surface hardness.

The surface region may be harder or softer than the bulk material as a result of several surface treatments. This difference in hardness may be accomplished deliberately in accordance with the requirements for the finished product. An example is the carburization process already mentioned, the mechanical effect of this treatment being that the surface layer becomes harder than the core.

The values for the Brinell hardness of some bulk materials are given below :

```
construction steel (0.1% C)                          105  kg/mm
CrNi steel (0.35% C - 4.5% Ni - 1.3% Cr)             240  kg/mm
aluminium                                          23-45  kg/mm
```

Surface roughness.

The surface morphology and texture of a metal are always
the result of various preparation procedures, machining
and polishing processes.
   There appears to be some evidence that there is a
connection between the surface roughness and the fatigue
strength.
   Surface roughness also has an influence on the
adherence of a sprayed metal, for example. Several optical
and mechanical methods are known for determining the
rugosity, but it is outside the scope of this chapter to
describe them. The degree of roughness can be determined
by comparing the unknown specimen with standard metal
samples of different types and degrees of roughness.
Inspection is normally carried out visually and by
palpitation.
   An indication of the roughness is given by the ratio
(roughness factor) between the value for the true surface
area and that for the mean surface area, both of which are
measurable quantities.

Chapter Two

Changes in the appearance of the skin of the façade

2.1 INTRODUCTION

The factors that determine the changes in appearance of
the skin of a façade, described in Chapter one, operate
simultaneously. In some cases the changes in appearance
are due only to deposits or liquids present on the surface or filling the pores; in other cases the changes are
due to degradation of the skin itself by chemical or
physical action. If degradation has proceeded too far,
cleaning alone will not be sufficient to re-establish an
acceptable appearance. The main factors determining the
changes are :

- The influence of air and rain, not including the
  solid particles present in air. This factor will
  continue to play a part after cleaning or restoration
  has been carried out. At that stage, it is sometimes
  possible to reduce its effect in the futuure by im-
  pregnation of the material, for instance.
- The influence of constituents that have migrated from
  the inside of the material. Before the surface is
  cleaned consideration should be given to arresting
  the source or stream of soluble materials.
- The influence of the adjacent materials. This is
  often an important factor during the building period
  and it should be eleminated before cleaning takes
  place.
- The influence of the soiling process itself, which is
  caused by solid material that is not soluble in
  water. This includes biological and non-biological
  deposits.

2.2 CHANGES CAUSED BY ATMOSPHERIC CONSTITUENTS

2.2.1. General description of the atmosphere

Many materials can react with the constituents of air and
rain und undergo changes in the presence of ultraviolet
light. Distinguishing between the external influence of
the atmosphere and deposition we have :

1. Dry deposition of the reactive gases (HNO_3, HCl, SO_2, NO_2, CO_2, H_2S and O_2) and of water-soluble solid materials such as seasalt.
2. Wet deposition of rainwater. This water can bring the dry-deposited products into solution. It can attack the façade material itself and can furnish the wet condition necessary for corrosion. It can also transport deposited products on the façade or even into the surface itself.

The flux F of dry deposition can be estimated from the equation :
$$F_d = V_d \times C_d$$
where V_d = dry deposition velocity
C_d = mass concentration in the air

As an example of how this formula can be used, an estimation for a country such as The Netherlands is given in the following table :

	atmospheric concen- tration ($\mu g\ m^{-3}$)	V ($m\ s^{-1}$)	flux ($\mu g\ m^{-2}.s^{-1}$)
O_2	300×10^6		
H_2O	$5-25 \times 10^6$	(depending on realtive humidity and temperature)	
CO_2	600×10^3		
solid particles (soil)	30-120		
seasalt coastal, 10 μm	300	5×10^{-2}	15
inland, 0.5 μm	2	5×10^{-5}	10^{-4}
SO_2	50	10^{-2}	0.5
NO_x	50*	10^{-2}	0.5
SO_4 (in aerosol particles of 0.1-1 μm)	10	5×10^{-5}	5×10^{-4}
HNO_3	5	5×10^{-2}	0.25
H_2SO_4 (in aerosol particles of 0.1-1μm)	1	5×10^{-5}	5×10^{-5}

*in streets with heavy traffic 150 μg m

Compared with the flux of dry deposition the flux of wet deposition is much smaller.

The flux of larger particulate matter also depends on the mass of the particles. Investigations have shown that for a certain locality the amount of solid particles that become attached to a parafined surface is proportional to the average particle content per m^3 and the average wind velocity expressed as the vector component in the direction perpendicular to the surface. The proportionality factor was found to be small: less than 1% of the particles reach the surface.

2.2.2 Weathering of concrete

2.2.2.1 Introduction
The four main components of the atmosphere that can react with the concrete skin are:

    water (vapour)   : 5-25 g per 1.3 kg (1 m^3)
    carbon dioxide   : 5-20x10^{-1} g per 1.3 kg (1 m^3)
    sulphur dioxide   : 2.5-15x10^{-5} g per 1.3 kg (1 m^3)
    chlorides   : 0.5-4x10^{-5} g per 1.3 kg (1 m^3)

As already mentioned, we can distinguish between the influence of the components as gases and when contained in rainwater.

2.2.2.2 The influence of water vapour - H_2O
Due to capillary condensation the moisture content of the concrete skin is strongly influenced by the relative humidity of the air. As the outer layer contains about 80% cement paste the pore content and therefore the moisture content of the skin is higher than that of the inner concrete when both are in equilibrium with the water vapour in the air.

Thus at 90% R.H. the skin can contain, for instance, 13% moisture and at 75 % R.H. about 11%. The moisture in this layer determines the rate of reaction with the other gases mentioned and also the permeability for gases. Thus the reaction rate of CO_2 is highest when the concrete is in equilibrium with air at about 60% R.H.

When the concrete dries out the drying process is often described by :

$$\frac{dW}{dt} = \frac{d}{dx} Dw \frac{dW}{dx} \qquad \text{where}$$

    W = evaporable water
    Dw = coefficient of moisture diffusion
    x = distance from the surface

The Dw depends on the water content and is also different for the skin and core of concrete.

When rain falls on wet concrete the equilibrium : tobermorite + water $\rightleftarrows$ lime + silica gel does not exist. Therefore the skin will be covered by a very thin but weak layer of silica gel. Underneath this layer there exists a dense carbonated layer which protects the concrete from further deterioration. This can be demonstrated by the method of sand-blasting when the soft outer layer and the dense layer can be detected after weathering has taken place in the first millimetre.

2.2.2.3 The influence of carbon dioxide - CO_2

When carbonation has taken place the density of the skin and the concrete becomes much higher. It is a barrier against the solution of water. Only at the outer skin the product of carbonation not present due to the formation of the soluble bicarbonate or gypsum is SO_2 is available.

The flux of CO_2 from the air is so high that a thick layer could be entirely carbonated if the diffusion rate in the concrete were high enough. CO_2 reacts with two components of the concrete : the lime and the ettringite.

$$CO_2 + Ca(OH)_2 \rightarrow CaCO_3 + H_2O \qquad (1)$$
$$3CO_2 + C_3A \cdot 3CaSO_4 \cdot 32H_2O \rightarrow 3CaCO_3 + 2Al(OH)_3 + \qquad (2)$$
$$3CaSO_4 \cdot 2H_2O + 23H_2O$$

In both cases water is formed. This water has to evaporate and in the meantime it will lower the gas diffusion of CO_2. Thus the carbonation rate is a function of the carbonation, the CO_2 permeability and the moisture content. If the concrete is already dense the carbonation will be limited to a depth of a few millimetres.

2.2.2.4 The influence of sulphur dioxide - SO_2

SO_2 can react with the lime and the oxygen from the air to form gypsum. This gypsum is able to furnish adherence for the dust particles. As H_2SO_4 is a much stronger acid than H_2CO_3, the SO_2 will expel the CO_2 from the outer layer.

The flux of SO_2 at the surface is so small however (30 gm/year) that if all the cement were to react only $\pm$ 0.1 mm of the top layer would be destroyed in the first year.

The flux of SO_2 deposition by the rain is less, even if the pH = 4.5. If 750 l rain fall in a vertical plane on 1 m^2 in a year, this quantity transports about 7.5 g SO_4 and 3.15 g NO_3 and this amount of acid can bring 7.7 g CaO into solution. Thus the influence is limited to the upper skin only.

However, acid rain is important because by washing out the first 0.1 mm it loosens from the surface the solid particles that have been deposited.

2.2.2.5 The influence of chlorides - Cl

Chlorides react with the C_3A of the cement to form a hydrate. The flux of chlorides near the coast is more than 100 times greater than elsewhere. Movement in the direction of the reinforcement is only possible by diffusion in water. If the water system in the pores is not continuous, the chlorides will not penetrate and the salts will stay at the surface. this can also be the case if the first 3 mm of the concrete has been made hydrophobic.

It can be said in conclusion that only the first millimetre in good dense concrete is a subject to weathering. Rain brings the top layer of the concrete into solution and is therefore able to clean soil particles from the surface.

The top layer to a depth of several μm becomes soft and will have a large specific surface. This is important for the adherence. Gypsum formed under the influence of SO_2 can also contribute to the adherence in places where the rain does not wash the façade.

If the cement content is very low weathering can take place further inside the concrete. This is shown in fig. A2.1 of appendix 2, where a concrete block (sand-gravel) was placed in (wet) sand for a period of 15 years. The exposed top was weaker to a depth of about 3 mm. The bottom cured in ideal condition.

2.2.3 Weathering of stones and bricks

Following a clear treatment recently proposed (2) the weathering of porous constructional materials, such as stones and bricks, can be considered as a kind of stress-corrosion process in which chemical attacks are combined with mechanical stresses.

Corrosion processes are, for example, the attack by rainwater or by condensation and the corrosive action of polluted atmospheres.

Some mechanical stresses may arise from internal mechanisms such as frost, salt crystallization and corrosion of iron cramps. Many others, on the contrary, are induced in the material by several external factors such as thermal expansion effects and stresses caused by cleaning and working techniques. Finishing techniques using, manually or mechanically, different tools (such as, for example, bushhammer, boaster and chisel) can produce different effects according to the type of material but, in any case, they can be considered as a potential damaging factor (3).

The alteration of stones and bricks generally proceeds from the external surface to the core of the material : even microscopic cracks and small increasing of the porous structure of the skin permit the ingress of agressive agents and of water, with a consequent acceleration of the alteration.

References

(1) Rossi-Doria P., (1985) Principles and applications of Pore Structural Characterization. Bristol p. 441.
(2) Torraca G., (1982) Porous Building materials ICCROM Rome.
(3) Alessandrini G. (1981) The conservation of stone, Int, Symp., Centro "Cesare Gnudi", Bologna, Part A, p. 139.

2.2.4 The weathering of wood

2.2.4.1 Introduction
The weathering of wood can be described as the
deterioration of the outermost skin of unprotected wood by
exposure to light and the weather. The effects are due to
a complex combination of chemical, mechanical and light
energies. It is usually said that slow erosion occurs,
i.e. 10 to 130 μm per year. But what is going on in that
first millimetre of the wooden skin? There are many
questions to be answered. The weathering process can be
divided into influences from outside the wood, the wood
itself, other circumstances and the effects of all these
phenomena.
   Thus degradation during external exposure is a result
of several influences.

2.2.4.2 Influences
   (a) Outside (environmetal) influences :
      - UV light: photochemical deterioration of the
        wood substance : colour change
      - rain and other climatological sources : wetting
        and drying, shrinking and swelling, grain
        raising, loosening of surface fibres, freezing
        and thawing
      - biological (fungal) attack : may collect, colour
        change
      - dirt : may contribute
      - temperature : drying if higher temperature etc.
      - oxygen : may contribute

   (b) Influences related to wood itself :
      - several wood species, including cellulose and
        lignin contents
      - cross sectional, radial and tangential plane
      - portion of early wood in the growth rings
      - pith or bark side
      - extractives, i.e. tannins, etc.

   (c) Other influences :
      - orientation north, east, etc.
      - vertical or other exposition
      - situation and architecture of the construction
        etc.

   However, weathering has little effect below the surface
of the wood. The electron microscope has detected mildew
in weathered European spruce that has penetrated through
ca. 20 cells and degradation of the wood substance (lignin
and celluloses) that occurs up to a depth of ca. 5-7 cells
(both in flat-grained performance). Cracking developed
over a depth of more than 20 cells (ca. 0.6 mm).

With the naked eye it is possible to see that smooth wood
surfaces have become rough, cracks develop and the colour
of the wood changes to yellow and brown, sometimes bluish.
A soft grey silvery colour appears on several wood species
when microorganisms ; mildew, for instance, are absent.

2.2.4.3 Chemical changes in weathered wood

These changes are a combination of the important effect of
light with other weathering factors. The following
distinguisable research results have been obtained so far:

- intercellular substance (principally lignin) has been
  lost mechanically (due to wetting-drying) ; the
  chemical structure has altered ; lignin decomposes
  and is subsequently more soluble,
- the grey layer consists of pure and degraded
  cellulose and lignin,
- intercellular substance (particular lignin) has been
  lost, first in rays and vessels, later in fibres
  (hard-woods) and tracheids (softwoods),
- cracking orientates spirally in cell walls and
  through pits,
- springwood degrades faster,
- the hygroscopicity of the wood increases, density decreases,
- cellulose appears to be considerably less affected ;
  only in the outermost surface it is affected more
  (for pine after a weathering period of ca. 20 years :
  unchanged wood, total cellulose 62.7%, lignin 27.8% ;
  in the weathered grey layer, cellulose 58.4%, lignin
  14.6% - a significant difference),
- the penetration of light is significantly different ;
  infrared penetrates more deeply than visible light,
  UV light penetration is negligible (0.016% of the
  visible light penetrates only as far as 2.5 mm),
- the photochemical process of colour change depends on
  the presence of oxygen,
- the grey colour is probably brought by iron salts :
  high humidity and temperature favour discoloration
  reactions.

Ultraviolet light.

Chemical deterioration is influenced greatly by the
wavelength of light, the most severe being that of
ultraviolet light. This results in photodegradation of
the wood close to the surface. All woods become an
overall soft grey : when accompanied by dark-coloured
spores and fungi mycelia the wood becomes a darker grey
and is often unsightly. Photodegradation takes place only
over a few cells (ca. 60 µm). The effects mentioned are
cumulative with other weathering factors.

The degradation of cellulose by light.

Cellulose is capable of absorbing some ultraviolet light.
It is believed that the acetal - chromophore components
are responsible for this. The C1-C2 bond (see fig. 2.1.)
in a glucose unit can be broken. However, no simple
sugars are found. Possibly due to lignin - cellulose
interaction during photodegradation, different product
develop from wood celluloses. Although this finding, and
that of the formation of free radicals during
photodegradation of cellulose (introduced by splitting of
the cellulose chains) are the result of research on
purified wood cellulose, similar processes will take place
in wood. Also it has been found that oxygen and water
vapour accelerate degradation in near-UV light (ca 388
millimicrons), while a higher temperature further
increases degradation. In this regard the difference
between pure cellulose and wood is larger, because studies
have shown a decrease in methoxyl content and also the
formation of lignin fragments (e.g. vanillin) under the
conditions in question.
    Other points with regard to UV reactions :

 - The absorption of UV light by wood is principally due
   to its absorption by lignin.
 - Some types of energy transfer from lignin to cellulose. Lignin functions as a photo sensitizer in the
   degradation of other wood substances (celluloses).
 - When the vessel wall of a poplar was examined after
   2 weeks of UV light exposure a large proportion of
   the cell-wall and bordered pits had been destroyed.
 - Photo-oxidation is important for wood degradation by
   2537 Å UV light. Lignin is subject to self-
   degradation as well, and is also a photosensitizer
   for cellulose oxidation.
 - The variation in wood components (cellulose, lignin
   and extractives) may be different in different wood
   species.
   Important types of wood are spring- and summerwood,
   heart- and sapwood and reactionwood. It is known
   that springwood degrades more than summerwood. There
   are two reasons why springwood degrades more than
   summerwood : the surface of springwood is relatively
   larger and the lignin content is higher in springwood.
 - Extractives can accelerate or retard photo degradation. A typical example of the latter is redwood,
   in which extractive migration causes the well-known
   dark, reddish-brown colour. When this is leached
   out and the surface becomes dry enough the grey
   layer appears.
 - Species effect. The difference in lignin etc.
   content is important. Also, the difference between

the lignin of hardwood and the lignin of softwood
results in a difference in degradation.

2.2.4.4 Cell structure changes.
Besides the important process of chemical degradation that
is greatly influenced by the wavelength of light (most
strongly by UV light) physical changes take place due to
wetting and drying cycles. Cracks are easily
distinguished. Fewer cracks appear in wood of lower
density than in species of higher density.

- A sharp swelling gradient occurs when rain makes
  contact with wood. Through rapid drying after a
  heavy rainshower (wetting) shrinkage cracks occur as
  a result of sharp tensile stress in the surface (the
  surface, densified by wetting is restrained by the
  dry core).
  The most severe strains develop in springwood.
- We know that the cell walls of tracheids (softwood)
  and fibres (hardwood) are built up of several layers.
  The middle layer of the secondary wall layer (the
  S2 layer) has a cellulose fibril direction of ca.
  15 degrees to the main direction (longitudinal axis)
  of the fibre.
  So-called diagonal weathering cracks (long, very
  narrow openings with sharp ends) are due to local
  concentrations of tensile stresses at right angles to
  the cellulose fibril direction of the S2 layer.

Several patterns of cracking are possible. Besides
boatformed cracks, Parallelepiped-shaped void spaces are
also discovered. They are caused by shear stresses along
the longitudinal axis of the fibres. Reasons for this and
the formation of diagonal cracks include a higher porosity
of the wood surface (see fig. 2.2.).
Weathering cracks that cannot be seen without
magnification develop in individual cell walls and between
adjacent cell walls. It is significant that these cracks
were not found in springwood. Sometimes rays may bring
about longitudinal cracking. A further possibility is
that cracks in the middle - layer substances occur as a
result of photo degradation of the lignin component.
Thus the development of stresses that cause cracking of
the surface is the result of a complex combination of loss
and alteration of wood components through photochemical
degradation, temperature and moisture changes, migration
of extractives, etc.
It is also found that porosity caused by cracking is
very different from that caused by soft-rot attack. The
cavities of soft-rot are more tubular (with abrupt
restricted ends). However, they also prefer the thick S2
layer and the well-known direction of the fibril angle ;
the shape is so different from weathering cracks that no

mistake is possible (fig. 2.3.).   Additionally, however
there are several fungi that bring about an increase in
moisture content, which again leads to further
degradation.

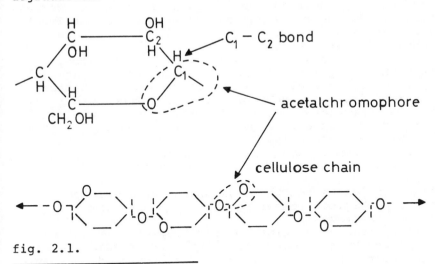

fig. 2.1.

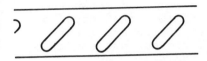

fig. 2.2.a.
Normal diagonal checks ;
increasing void area (a longitudinal
wall of a late springwood tracheid).

fig. 2.2.b.
Unusual shape of weathering checks,
strands of (cellulose etc.) wood-
substance of nearly right angles
to the long axis of the tracheid.

fig. 3.
The typical voids formed in the
S2-layer by attack of soft fungi.

2.2.5 Weathering of glass

2.2.5.1 Introduction
When glass is exposed to atmospheric conditions, it more
or less undergoes degradation. This degradation is termed
weathering. In theory, glass can deteriorate in at least
three degradation modes : by mechanical, chemical and
photochemical degradation.

Mechanical degradation of glass includes abrasion,
fracture and fatigue by atmospheric agencies such as hail,
sand particles, thermal stresses and windload. Chemical
degradation of glass, known as glass corrosion, occurs by
reaction of the glass surface with atmospheric water in
both liquid (condensation, precipitation) and vapour form.

Some glasses, depending on the composition, change
colour upon exposure to (solar) UV radiation by
photochemical reaction. This phenomenon is called
solarization.

The mechanical strength of glass can be increased by
thermal and chemical tempering methods during the
manufacturing process.

The solarization of glass mostly becomes apparent only
upon exposure to extremely high levels of irradiation.

Glass corrosion, however, cannot be prevented by simple
means. Aqueous corrosion of glass has been studied for
both scientific and practical purposes for a great many
years. Recent publications about the chemical durability
of glass and theoretical treatments of glass corrosion
can be found in references 1, 2, 7, 9 and 10.

The object is to give a short review of the corrosion
mechanisms of glass in view of the application of glass in
windows and solar energy systems. Glass is an important
candidate for use as cover glazing for flate-plate
collectors and as encapsulation for photovoltaic cells (4,
6). Furthermore, several glasses are under consideration
as superstrates for solar mirrors (4, 5).

Although it is known that other glass formulations like
borosilicate and aluminosilicate glasses are more durable,
the application of common sodalime-silicate glass is
preferred for economic reasons. Therefore, attention is
focussed on the corrosion of soda-lime glass. Results of
recent theoretical studies and typical weathering
experiments are summarized.

2.2.5.2 Mechanism of glass corrosion
The interaction of glass with liquid water is known as
aqueous corrosion. The chemical reactions that occur at
the glass surface depend on a number of factors : the
composition of the glass, the pH of the aqueous solution,
the temperature, etc. It has been shown by many
investigations (1, 2, 10) that the aqueous corrosion of
common soda-lime-silicate glass can be represented by the
following, simplified two-stage model :

30

In neutral or acidic solutions with pH < 9 water penetrates into the glass network and alkali (Na^+) ions in the
glass are replaced by hydrogen ions (fig. 2.2.5.2. 1) :

$$Si - O\ Na + H_2O \rightarrow Si - OH + Na\ OH \qquad (1)$$

Outward diffusion of the alkali ions leads to a
dealkalized surface layer.

Under static solution conditions (cf. 2.2.6.3), i.e.
when the leaching solution is not replenished, the pH of
the liquid increases by the accumulation of the Na -ions.
In basic solution with pH greater than 9 the silicate
structure itself is attacked by hydroxyl ions which break
down the siloxane bonds :

$$Si - O - Si + OH^- \rightarrow Si - OH + Si - O^- \qquad (2)$$

Silica can then be extracted and total dissolution of the
surface will occur.

It is generally accepted that the nature of the leached
surface film, which is also known as alkali-deficient,
silica rich or hydrogen glass is very important in
understanding the durability of glasses. Hench (2) has
identified six types of surface conditions which are
characteristic of a silicate glass at any time in its
identified six types of surface conditions which are
characteristic of a silicate glass at any time in its
history. He concluded that durable glasses almost always
develop a stable surface film which has a higher
concentration of network formers (SiO_2) than the bulk
composition.

The kinetics of the corrosion of the glass surface are
rather complex. Smets (10) concluded on the basis of
experimental data that the rate-determining step in the
aqueous leaching of glass in solutions with pH < 9 is the
inward diffusion of molecular water. The stage 1 process
of alkali leaching is diffusion-controlled with a $t^{\frac{1}{2}}$
dependence (1). Stage 2 of network dissolution is
controlled by an interface reaction with linear (t^1) time
dependence (1). The transition from stage 1 to stage 2
reactions can, in principle, occur over a short time
interval. The transition rate, however, depends mainly on
the time required for the solution pH to reach the high
alkaline range and is therefore very sensitive to local
concentration effects.

In general the temperature dependence of the corrosion
rate is exponential : $\exp(-E_A/RT)$. This implies that the
corrosion becomes more rapid at increasing temperatures.
E is the energy needed to activate both the alkali
leaching and the network dissolution process ; $E_A \sim$
$80 kJ/mole$ for H_2O (7).

An important parameter with respect to the corrosion
rate of glass surfaces has been defined by Hench (1) : the

localized surface area to solution volume, SA/V (cm^{-1}).
Two objects in intimate contact have SA/V values greater
than 100 cm^{-1} whereas a free surface has a SA/V value much
less than unity. At high SA/V conditions surface attack
by water becomes greatly accelerated and only short times
are needed to show surface deterioration as in the case of
wet storage of window glasses.

2.2.5.3 Exposure conditions
Two types of exposure to liquid water have been
distinguished (3, 9) : dynamic and static. Under dynamic
conditions water in contact with the glass surface will
extract only alkali (Na$^+$) ions according to eq. (1).
    Reaction (2) will not take place because the water
carrying the reaction products is continuously replenished, preventing the pH reaching high alkaline values.
    However, when the glass surface is exposed to static
conditions the leaching products collect and the pH of the
solution will eventually reach values higher than the
critical value for network dissolution (pH > 9). Consequently the silicate structure is damaged according to
reaction (2).
    Weathering of glass by water vapour involves adsorption
of water vapour by the glass surface and the leaching process of eq. (1). The reaction products (NaOH) crystallize
on the surface and, since they are very hygroscopic, will
adsorb water. The glass surface will then be attacked in
the local areas where crystals have formed, leading to a
pitted surface.
    The rate of weathering due to atmospheric attack is to
a large extent a function of local climate and systemic
conditions and cannot be described by a simple model.
Actual weathering experiments are needed to characterize
the rate kinetics.

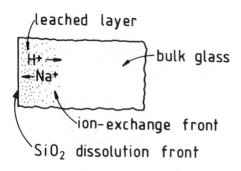

leached layer

H$^+$ →

← Na$^+$

bulk glass

ion-exchange front

SiO$_2$ dissolution front

fig. 2.2.5.2. 1.

References

(1) Hench, L.L. and Clark, D.E. (1978) Physical chemistry of glass surfaces. J. Non-Cryst. Solids 28, 83.
(2) Doremus, R.H. (Acad.Press 1979) Chemical durability of glass. Treatise on materials Science and Technology, 17,41.
(3) Franz, H. (1980) Durability and corrosion of silicate glass surface. J. Non-Cryst. Solids 42, 529.
(4) Lind, M.A. and Hartman, J.S. (1980) Natural aging of soda-lime-silicate glass in a semi-arid environment. Solar Energy Mat., 3, 82.
(5) Shelby, J.E., Vitko, J. and Pantano, C.G. (1980) Weathering of glasses for solar applications. Solar Energy Mat., 3, 97.
(6) Bouquet, F.L. (1981) Glass as encapsulation forlow-cost photovoltaic solar arrays. J. Solar Energy Engr., 103, 221.
(7) Dunken, H.M. (1981) Physikalischen Chemie der Glas-oberflächen. VEB Deutscher Verlag für Grundstoff-industrie, Leipzig.
(8) Norville, H.S. and Minors, J.E. (1985) Strength of weathered window glass. Am. Ceram. Soc. Bull., 64 (11), 1467.
(9) Newton, R.G. (1985) The durability of glass - a review. Glass Technology 26, 21.
(10) Smets, B.M.J. and Tholen, M.G.W. (1985) The pH dependence of the aqueous corrosion of glass. Physics Chem. Glasses, 26, 60.

2.2.6 The weathering of plastic building components

The plastic building components placed in the external envelope of a structure are subjected to various climatic stresses. The environment involves different parameters, principally : solar radiation, water (in solid, liquid or gaseous form), air as well as various pollutants. To this list must be added the biological parameters (micro-organisms, plants, animals).

All these parameters will act together through chemical, physical, biological processes (mechanical, thermal loading, abrasion....).

It is essential to study the overall behaviour in the light of the various interactions (favourable or unfavourable) which the factors can exert. The actions and mechanics of one or several parameters exerted simultaneously are very complex. The effects of stresses will be marked by modification on a microscopic and macroscopic scale. there will be modification in composition or structure; also changes in aspect, dimensional variations etc.

The soiling observed on cladding and other polymer-based materials results from the action of different pollutants, generally in association with water.

The effect of pollutants will be exerted by the modification in aspect as well material changes. As for the other parameters, the direct and the indirect effects can be distinguished. As an example of the indirect effect of pollutants, it will be observed that soiling deposited on the surface will increase the absorption factor of the material in relation to solar radiation (infra-red radiation) which will bring about an increase in temperature with all the consequencees this implies from the point of view of chemical actions (acceleration of reactions), physical actions (evaporation of volatile constituents, differential expansion....).

Effect of climatic ageing on surface properties

As already stated, the effect of pollutants on soiling of building components cannot be studied without taking into account the mechanisms of actions and effects on the polymer and on the other climatic ageing parameters. These include the actions and effects on the polymer and on the other ingredients in the formulation (fillers, pigments, plasticizers, various reinforcing agents and admixtures..).

As regards polymers, the principal mechanisms are photo-oxidation and thermo-oxidation. Photo-oxidation is due to the fraction of the solar radiation spectrum, below about 400 nanometres (ultraviolet). this photo-oxidation brings about a break in chains and other secondary reactions. The mechanical effect is generally micro-

34

cracking which, according to the type of formulation and
its photmetrical characteristics, is limited to the
surface or can penetrate greater thickness.   Once this
break in the chain and the microcracking have occured,
various phenomena may follow in sequence.
     Since the surface state is affected also by the surface
tension, water absorption will increase appreciably.
From then on the pollutants dissolved in the water will
penetrate into the cracks and settle there.   Once the
first soiling is formed, further soiling will follow.
     Other phenomena may be the cause of soiling fixing in a
given place, for example migration of certain constituents
of low molecular weight contained in the formulation.
     Biological agents can also intervene in the soiling
process.   They act according to several processes, the two
main factors being :

- assimilation of the material or of certain compounds
  of low molecular weight,
- actions due to the products of micro-organism metabolism (action by chemical processing or by fixing of
  pollutants).

In order to develop, micro-organisms must have food
available which can be assimilated by the species.   This
is the case of :

- decomposition products in the polymer chain,
- products of low molecular weight present in the formulation at the beginning of the process (plasti-
  cizer, internal demoulding agent, fillers....),
- other pollutants already fixed on the material.

The biological agents or products of their metabolism
themselves contribute to soiling.

Elimination of soiling

Before putting into application any process to remove
soiling, it will be useful to undertake some preliminary
investigation such as :

- examination of the origin and the nature of the
  soiling
- examination of the state of surface degradation of
  the material, independently of the soiling,
- examination of the side effects of a process to remove soiling.

Soiling can be removed either by dissolving or dispersing,
or using a more or less abrasive method.   If need be,
these methods can be combined when preferable.   The
cleaning liquid can contain agents which will react

chemically with the soiling of fungus-destroying agents, stopping the development of biological agents.

After the treatment, it will be useful to examine the state of the skin and its potential capacity to attract fresh soiling. This will also be the time to raise the problem of any modification to the surface, either by mechanical treatment or application of a protective coating.

Conclusions

The development of soiling on plastic building components subjected to the climatic environment, cannot be dissociated from the general process of ageing.

It is preferable to choose materials ensuring satisfactory intrinsic resistance to ageing and with a surface tension and the possibility of setting up and maintaining electrostatic charges can be adapted by means of external modification or by incorporating additives in the formulation of the plastic components.

The inesthetic effect of soiling will be attenuated when it is homogeneous all over the component. It is essential to avoid the formation of areas of differtial washing. For this, steps must be taken at the design stage of the different components placed in walls.

2.2.7 The weathering of metals

2.2.7.1 Introduction

The weathering of metals is a complicated series of
events, including contamination of the surface,
degradation of the metal surface or a thin surface film by
wear, and further deterioration of the metal by one of
various forms of corrosion. Apart from exceptional
circumstances such as strong winds in coastal areas with
sand dunes or in desert like regions, it is corrosion that
is the major problem for metal façades. We will therefore
concentrate on this phenomenon.

In the ambient atmosphere we can distinguish between
gaseous components, either as contaminants or as major
components, and solid particles, either dissolved in water
or present as an aerosol. Three principally different
types of atmosphere exist : marine, urban and industrial.
As far as corrosion is concerned large areas of rural
regions can nowadays be considered as industrial because
of extensive contamination by manure.

Gaseous components include the following : oxygen,
ozone, water, carbon dioxide, sulphur dioxide, hydrogen
sulphide, ammonia and other acid-forming gases such as
nitrogen oxides and chlorine. Very often chlorides and
dust particles can also be observed in the atmosphere (of
both organic and inorganic origin) next to industrial
dust, with a composition that very much depends on the
local situation.

In general the soiling of façades increases the changes
for an increased corrosion rate. During periods with high
relative humidity dirt is precipitated onto the surface.
When the relative humidity decreases again, the
contamination dries out and remains as solid dry particles
at the surface. In this way, especially when the nights
are cold and the days warm and dry, the contamination of
façades can increase rapidly. Cleaning is then required
several times per year to prevent unacceptable corrosion.
Therefore places with good accessibility for rainwater
perform better than more or less occluded places.
Moreover, the accumulation of moisture within crevices
will often result in accelerated corrosion. A uniform
description of the results of soiling cannot be given,
since different materials will behave differently under
varying atmospheric conditions.

As an example low-alloy construction steel can be
considered. In dry air this material is protected by a
thin homogeneous oxide layer. However, if the relative
humidity surmounts a value of 50-60% at room temperature a
film of moisture will form, resulting in an increased
corrosion rate. The presence of salt particles near the
sea or ocean can increase the impetus for moisture to
condense and thereby bring about deterioration in the
morphology of the corrosion products. The already poorly

protective nature of the surface layer will then disappear
completely. The same effect can be observed in industrial
regions for sulphur dioxide and dust particles.

2.2.7.2 Weathering and corrosion
Weathering in the strict sense plays a part for layers of
lacquer on aluminium alloys used in façades. However, as
far as metals are concerned corrosion in its various forms
is a more appropriate description of the process of deterioration due to the atmopheric conditions. For façades
the following forms of corrosion are most relevant :

 (a) General corrosion, where the metal corrodes more or
     less uniformly over the total surface.
 (b) Galvanic or contact corrosion, when two metals with
     different nobility are in contact. The less noble
     metal will corrode at a higher rate as a result of
     electrical contact with the other metal.
 (c) Filiform corrosion, where under a thin paint or
     lacquer layer corrosion proceeds in such a way that
     the corrosion product appears as "fibres".
 (d) Crevice corrosion, where differential aeration is
     responsible for the enhanced corrosion within the
     crevice.
 (e) Pitting, with a simular mechanism to crevice
     corrosion. The chloride content of the solution is
     very important.
 (f) Intercrystalline corrosion, where the metal is at-
     tacked along the boundaries of the crystallites.
 (g) Stress corrosion cracking, due to the combined
     action of tensile stresses and corrosion. Both
     transcrystalline and intercrystalline cracks can
     result.

2.2.7.3 The character of corrosion reactions

Metal ions are formed as e.g. for iron according to the
reaction : $Fe \rightarrow Fe^{3+} + 3e$, while electrons are also
produced. It is clear that this reaction can only proceed
if further electrons are consumed elsewhere. Two
reactions can be considered as the most important :

$2H^+ + 2e \rightarrow H_2$, acid corrosion, and
$O_2 + 2H_2O + 4e \rightarrow 4OH^-$ oxygen corrosion.

Both cathodic reactions show clearly that acid rain is
unfavourable, but it is also evident that oxygen and water
play a part. The reactions given here are simplified. For
the low-alloy steel material already discussed these
should be given as :

38

$$Fe + H_2O \rightarrow FeOH^+ + H^+ + 2e$$

anodic reaction

$$Fe(OH)^+ + H_2O \rightarrow FeOOH + 2H^+ + e$$

$$3FeOOH + e \rightarrow Fe_3O_4 + H_2O + OH^-$$

cathodic reaction

$$Fe_3O_4 + 1/4\ O_2 + 3/2\ H_2O \rightarrow 3FeOOH$$

See also fig. 2.2.7.3. 1.

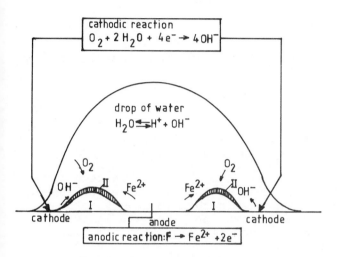

fig. 2.2.7.3 1.

Apart from acidifying the electrolyte SO_2 does influence the reaction by changing the cathodic reaction to :

$$SO_2 + O_2 + 2e \rightarrow SO_4^{2-}$$

A faster reaction, while the sulfate ions contribute to an increased conductivity of the electrolyte, thereby also increasing the corrosion rate.

Stainless steels and aluminium alloys, the main classes of metallic materials for façades, normally are covered with a passive layer. This means that these materials are less sensitive to general corrosion, but on the other hand display an increased tendency for pitting especially in chloride-containing surroundings (see fig. 2.2.7.3. 2).

This form of corrosion is initiated by the formation of small pinholes. These small holes originate especially if chloride salts precipitate at imperfect sites in the thin passive layer. These imperfections may result from contamination.

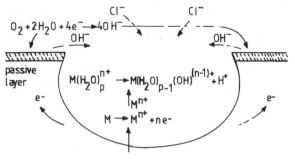

fig. 2.2.7.3. 2

Inside the pit the electrolyte acidifies according to the corrosion reaction:

$$Me + H_2O \rightarrow MOH^+ + H^+ + 2e$$

Also because Cl^- ions diffuse into the pit to preserve electroneutrality, thus increasing the electrical conductivity of the solution within the pit, repassivation of the surface within the pit is impossible. Due to the extended outer surface where the cathodic reaction takes place the pitwalls will then corrode at increased speed. Especially for stainless steels this is a very serious problem. For aluminium alloys the pitgrowth is much slower, due to the poor conductivity of the remaining passive layer outside the pit. In this way the cathodic reaction is slowed down, with a matching slower anodic reaction, since the total anodic current and the total cathodic current must be equal in the stationary state.

2.2.7.4 Corrosion and protection against corrosion for various metals

In general three different approaches either apart or in combination should be followed to prevent undue damage :

    (a) An optimal materials choice in the design stage of a building.
    (b) Application of protective coatings.
    (c) Appropriate maintenance and cleaning.

Therefore the discussion of various metals will be complex and incorporate aspects of all three approaches.
    The following materials will be discussed :

    - aluminium (-alloys)
    - stainless steels
    - weathering steel.

Aluminium

The rate of corrosion reactions is dependent on climate, orientation of the building and pollution of the ambient. Aluminium is attacked rapidly by alkaline solution, by hydrogenhalogenides, dissolved heavy metals and salts of the halogens. The roughness of the surface is also an important factor, because dirt and moisture/water will attach better to rough surface.

These factors will implicitly be discussed in the following, where the various forms of corrosion of aluminium-(alloys) as far as these are relevant for façades are treated.

Pitting

As already mentioned, the passive layer normally present on aluminium renders it sensitive to pitting corrosion. Chlorides in combination with a higher relative humidity induce pitting. Near the coast salts from the seawater precipitate on the metal surface, attract water and thus form a electrolyte, which provides the possibility for enhanced corrosion. Also in urban areas chlorides from nearby chemical plants and refuse incinerators, where also PVC's are burned , play a role. Also window cleaning should not be done with chlorine-containing solutions especially when the aluminium alloy was coated or anodized.

Also other components of the polluted atmosphere enhance pitting (like general corrosion). These are CO, CO_2, SO_2, C_xH_y, NO_x, NH_3, HF, and O_3. Some of these have a major influence on the acidity of the atmosphere, like SO_2, NO_x, and NH_3. A part of the SO_2 reacts to sulphuric acid in a humid atmosphere. In a similar way NO_x forms nitric acid.

By application of a thicker anodized layer up to 30 micrometer, attack can be reduced. In a non-polluted atmosphere anodizing is hardly necessary. In fig. 2.2.7.4. 1 the average results of expositions of various aluminium alloys to different atmospheres has been given.

Average pit depth in micrometer

	1 year	2 years	5 years	10 years
Marine atmosphere	8 – 30	18 – 36	22 – 40	24 – 40
Industrial atmosphere	22 – 56	42 – 60	46 – 74	46 – 74

For some of the alloys in figure 2.2.7.4. 1. additions of manganese and magnesium in combination with silicium increased the corrosion resistance. The thickness of the anodizing layer was 15-25 micrometer. From this figure it is clear that the corrosion process stopped after 5 years. This is the result of the precipation of solid corrosion-products in the pit. The protective action of this material can only persist if it is not dissolved in an aggressive solution. Therefore no aggressive detergents should be used, or the anodizing layer should be thicker.

Temperature fluctuations (difference between day and night, condensation of moisture), also influence the depth of the resulting pits.

Because of the strong influence of SO_2 and NO_x (e.g. in exhaust gases), on this type of corrosion, it is unwise to use anodized aluminium alloys for the inside of porches.

Aluminium in contact with concrete, often shows pitting corrosion under the influence of chlorides from the accelerators within the concrete.

Galvanic corrosion

Also this form of corrosion is promoted by a humid polluted atmosphere. The less noble metal will corrode at a higher rate, when in contact with a more nobel metal. This is also true for different alloys of aluminium in contact with each other. In figure 2.2.7.4. 1 the % of corroded surface area due to galvanic action for untreated aluminium in contact with various other metals and alloys has been given for two different atmospheres. It is clear that the corrosion is worse for combination with more noble metals and for an atmosphere with more chlorides and a higher degree of humidity. Therefore it is recommended to use isolation between aluminium and most of these metals and alloys.

Also combination of aluminium with brass, as is often done in sunshades and ventilation grids, leads to problems. Nylon is then an alternative material.

Metallic impurities (steel particles from drilling and blasting) must be moved from the surface for the same reasons. When using screws and bolts a suitable alloy (aluminium based or at least stainless steel), or an insulating ring must be used. In the proximity of rail traffic (up to 200 meters) aluminium façades suffer in a humid atmosphere very much from the deposition of copper particles stemming from the overheadlines. In a very humid atmosphere direct contact between wood and aluminium alloys must also be prevented, because wood impregnated with copper or mercury salts could be the cause of galvanic corrosion.

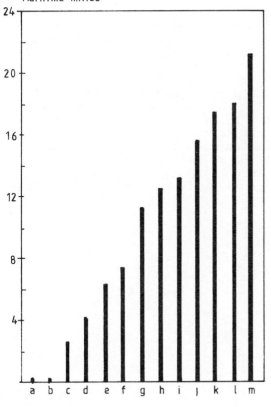

a : Zinc   h : Gold
b : Cadimium   i : Alloy
c : Steel   j : Stainless
d : 67%Ni. 30%Cu.   : Steel 16%Cr.
e : Nickel   k : Silver
f : Stainless Steel 18%Cr, 9%Ni   l : Iron
g : Lead   m : Copper

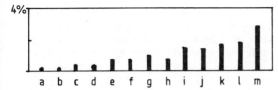

Corrosion contact of Aluminium in contact with other metals

fig. 2.2.7.4. 1

43

Filiform corrosion

This form of corrosion is often observed for lacquered
aluminium. Corrosion propagates starting from one point
producing "filiformed" corrosion products as shown in
figure 2.2.7.4. 2.
    This form of attack is the result of a poor or too thin
layer of the protective lacquer of paint. It is most
often observed at sharp edges and near pores or damage of
the protective layer. At a relative humidity of over 60%,
the deposit of dirt and salt particles promote this form
of attack. Filiform corrosion is seldomly observed at
places readily accessible for rainwater. A good pre-
treatment and an optimal lacquer application can prevent
this form of corrosion especially in combination with good
design of the façade structure.

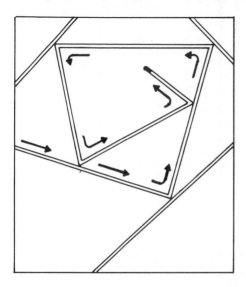

fig. 2.2.7.4. 2

General corrosion

When aluminium corrodes more or less homogeneously over
the surface, a white corrosion product remains on the surface. This product consists mainly of aluminium oxide.
For unprotected aluminium alloys this form of corrosion
initially proceeds at a considerable rate, but after some
years the rate decreases. The surface takes on a dead and
grey appearance. Pollution has been shown to have a negative influence. Horizontal surfaces exposed to rain suffer particularly from general corrosion. Here, too, attack is more severe in an atmosphere with a high relative

44

humidity. In this case (for non-polluted air above 66%) a very thin moisture film will be present on the surface acting as electrolyte. For a rough metal surface and if dirt and salts deposits are present this value is lower. It is absolute necessary in those cases to clean the façades regularly.

Thereby it should not be forgotten that :
The cleaning solution should not be alkaline, thus ammonia and soda cannot be used. Neither can solutions containing hydrochloric or phosphoric acid be used. The pH of these solutions must be   between 5 and 8.

If parts of aluminium façades during construction come into contact with fresh not hardened concrete/cement they can even in anodized form be attacked due to the high alkalinity of the mortar. Stains should be removed immediately, whereafter rinsing with pure water is required.

Crevice corrosion

Also crevice corrosion can occur with aluminium constructions. Only for very narrow crevices of 0.05 to 0.2 mm this problem exists, if moisture intrudes the crevice. The oxygen within the crevice is used up as a result of the cathodic reaction. Due to the persisting corrosion reaction within the crevice while the cathodic reaction takes place at the much more extended outer surface, the pH decreases within the crevice, while chloride ions diffuse into the crevice to preserve electroneutrality. The passive layer within the crevice is broken down and due to the high acidity and high chloride ion concentration no repassivation will occur. This mechanism resembles the pitting mechanism. Attack can be relatively fast, due to the high current density at the small crevice inner surface. Therefore crevices in façade constructions, should, if they cannot be avoided by optimal design, at least be sealed efficiently.

For the same reason aluminium sheets should before fastening to the façade not be stacked up in a humid environment.

Summary for aluminium

We have seen various aspects of protecting aluminium façades from unnecessary and unwanted corrosion. Apart from good designing and good maintenance measures, protective layers play an important role. This item has only been dealt with in a restricted way. We have seen the passive film and the thicker anodized layers on aluminium. However, anodized layers may also be coloured and sealed, to improve both optical and corrosion properties. Apart from these layers also chemical treatments, like chromatizing and phosphatizing, may improve the resistance to especially pitting corrosion.

Stainless steel

The resistance of steel against atmospheric corrosion is
increased considerably by the addition of chromium. At
least 12% Cr is required to obtain a passive film of
Cr_2O_3. Addition of nickel further increases the possibilities for application of the steel. We are then dealing
with the stainless steels. For application in façades
mainly AISI 304, 316 and 321 come into consideration. Addition of other elements like titanium, molybdenum, copper
and manganese in minor quantities also improves the properties.
   As with aluminium the resistance to corrosion of the
stainless steels is better for smooth surfaces with a
homogeneous composition. Deposits of dirt and alien metal
particles can cause local corrosion phenomena (contact or
galvanic corrosion, and crevice corrosion).
   Stainless steel can also suffer from intercrystalline
corrosion, especially in an acid environment. This form
of corrosion we have not discussed for aluminium alloys,
because there it is more rare especially in façades. However for stainless steels in the direct surrounding of
welds, where the temperature during welding can have
reached values between 450 and 850°C, chromiumcarbides may
have precipitated at the crystalboundaries. This results
in a local depletion of chromium immediately next to the
crystal boundaries, with intercrystalline corrosion in the
end. this intercrystalline corrosion in fact is a special
form of galvanic corrosion at the crystal boundaries. The
stainless steels AISI 304 and 316 have an average carbon
content not above 0.07%. Therefore they are, at least in
plate form with a thickness below 6 mm, not extremely
sensitive to this form of corrosion, especially if an
optimal welding procedure has been followed.
The third material AISI 321 is not sensitive at all,
because the carbon has been bonded to titanium, thus
avoiding the precipitation of chromiumcarbides even after
sensitization by welding.

The sensitivity to pitting corrosion is in principle the
same as for aluminium alloys. But as we have seen before,
the resulting damage is more extensive for stainless steel
than for aluminium, due to the high electrical
conductivity of the oxide film around the pit mouth. Also
here the chloride content is the most important factor.
Deposits of dirt can also be responsible for the
initiation of pitting. Good cleaning of the stainless
steel façade is therefore required. Because AISI 316
contains some molybdenum it is less sensitive to pitting
than the two other stainless steels.

Stress corrosion cracking was not discussed for the aluminium alloys, because this form of corrosion will only
become apparent when the construction suffers from stresses.   In buildings this will therefore mostly occur in the
constructive elements and less in the façades, and therefore more often in stainless steel.   All types of stainless steel suffer from this type of corrosion.   Below 50°C
the effects are less severe.   Chloride ions also here play
a negative role.   A higher nickel content decreases the
sensitivity to a certain extent.
    From these data it is clear that application of stain-
less steels in cramp-ions especially in a marine atmosphere will be unwise.
    For these special purposes alloys with a higher content
of additions are better suitable, like :
X10CrNiMoTi 1810 (Werkstoff.Nr. 4751) and
X10CrNiMoNb 1810 (Werkstoff.Nr. 4580).
When stainless steels are used in conjunction with Al
façades, a good electrical isolation must be applied to
prevent the façade from galvanic corrosion.   Screws, bolts
or rivets of stainless steel can be used in combination
with aluminium alloy plates because the surface of the
aluminium plates is big compared with the fasteners, so
that the anodic current density and therefore the corrosion rate at the aluminium surface will be negligible.

Crevice corrosion can occur as for aluminium if moisture
enters crevices.   A high Cr and a high Mo content of the
stainless steel will have a positive effect.

Weathering steel

We have seen that the composition of steels has a large
influence on the rate of the atmospheric corrosion in the
case of the stainless steels.   This however is also more
generally true.   In particular the presence of relatively
small amounts of Cu, Cr and Ni (thus for Cr and Ni in much
smaller amounts than for SS) can also be benificial.
    A general overview is given in fig. 2.2.7.4. 3.
    These steels are used under the name weathering steels,
because they can be applied without any additional corrosion protection, due to the development of a dense layer
of corrosion products, with a good adherence to the steel.
    Also due to their improved strength these steels can be
applied for bridges, steel constructions in industry and
also in architecture.
    However, if these steels are applied in architecture in
a climate where big differences between the temperature
and consequently the humidity during the night and during
daytime exist, they may develop a less protective layer.
This is e.g. relevant for all regions near the sea, where
the high chloride content of the atmosphere is an extra
negative factor, making these steels less promising com-

47

pared with mild steel. The major problem will then not be the loss of stability due to corrosion, but the pollution of other parts of the building by the brownish corrosion products of the steel.

Painting or structural architectural solutions will be required then. Thereby the advantages of these materials: lower maintenance costs, a beautiful colouring and smaller weight of the structure due to the higher strength, somewhat decrease.

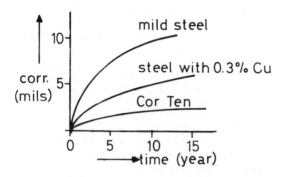

Composition of some types of weathering steel

Steel	C	Mn	P	S	Si	Cu	Ni	Cr
mild steel	0.16	0.6	0.01	0.03	0.01	0.01	0.01	0.03
Cor Ten A⁺	0.09	0.5	0.07	0.03	0.52	0.31	0.40	0.74
Cor Ten B	0.15	1.3	0.01	0.03	0.26	0.31	–	0.44
Mayari R⁺⁺	0.10	0.7	0.10	0.03	0.28	0.35	0.65	0.55

⁺ U.S. Steel
⁺⁺ Bethlehem Steel

fig. 2.2.7.4. 3.

References

1. Ergang, R. and Rockel, M.B. (1975) Werkstoffe und Korrosion, 26, 1, 36.
2. Wernick, S., Pinner, R., Zurbruegg, E. and Weiner, R. (1977) Eugen G. Leuze Verlag D. 7968 Saulgau, Zweite Auflage
3. Vargel, C. (1979) Le Comportement de l'aluminium et de ses Alliages, Paris.

2.3 CHANGES OF ASPECTS OF FACADES DUE TO PARTICULAR SUBSTANCES PRESENT IN BUILDING MATERIALS

2.3.1 Introduction

The number of possible processes that could be included in this chapter is considerable. Therefore a restriction is imposed that materials considered before-hand to be of little durability are anyway unsuitable to resist the severe influences to which façades are usually exposed. Also, those building materials are excluded where failure has occured during manufacture : for instance, light and dark areas on smooth concrete surfaces, which are a result of high and low W/C ratios in the outermost concrete skin.

Only visible material alterations such as decomposition and discoloration as well as those of a physical nature are considered. These alterations occur in the exterior and interior surfaces of the materials, where water plays an important part in the molecular metabolism.

Although the façade materials may be properly manufactured, the following almost spontaneous visible material alterations can change the surface appearance.

2.3.2 Concrete

Lime and alkaline salts dissolved in water can be transported to the surface of the concrete, where they will be deposited and become visible as efflorescence. The formation on concrete of efflorescence of lime which reacts with CO_2 from the air to form $CaCO_3$ is of great importance for the appearance of a façade. Only when efflorescence is deposited irregularly on the surface, is it judged inconvenient. There is a strong possibility that $CaCO_3$ efflorescence will disappear with the rainwater, because the $Ca(HCO_3)_2$ formed is a salt of good solubility.

When the colour of a new concrete façade is assessed from reflectance measurements, in many cases it is noticed that the reflectance increases during the first year and afterwards decreases as a result of the formation and later dissolution of the efflorescence.

Efflorescence is also responsible for the effect that the colour of steam-cured concrete is usually lighter than the colour of normally cured concrete.

With regard to the soiling aspect of the façade, the committee is of the opinion that no separate discussion is needed here of the mechanism of efflorescence formation.

A further point is that impurities in the aggregate can cause stains or eruptions. Bivalent iron compounds in the aggregate can cause discolouration on the surface, visible as rust stains.

Rust stains or even cracks can become visible due to the corrosion of the reinforcing steel. This can be caused by the penetration of CO_2 from the air, which neutralizes the protective effect of the $Ca(OH)_2$ ions against rusting of the reinforcing steel. Also, chlorides can stimulate this steel corrosion.

However, this phenomenon is not to be anticipated in a good quality concrete of sufficient thickness covering the appropriate quantity of steel reinforcement.

2.3.3 Bricks

A deposition of lime and alkaline salts can become visible on the surface of bricks. This efflorescence originates from the inner part of the brick and is transported with water to the surface. With this efflorescence there is a strong possibility that it will be dissolved in rain-water and disappear.

Bricks can also develop yellow and green stains resulting from vanadium salts. The vanadium salts originate in the raw material used to manufacture certain bricks. These local stains can be found on red, buff or white bricks.

Another phenomenon is that of eruptions on brickwork, which originate from rather large particles of CaO. These particles react with water to form $Ca(OH)_2$, which involves a volume enlargement and causes the eruptions.

2.3.4 Natural stone

On natural stone discoloration can appear in the form of iron stains. This is due to the presence of pyrites (FeS_2), which can be converted into iron compounds which are soluble in water and can be deposited as brown or black stains. For instance on white marble they are already visible when they are present just below the surface.

2.4 FACADE ALTERATIONS CAUSED BY ADJACENT AND ADDITIONAL MATERIALS

2.4.1 Introduction

In the designalization of a façade, materials are required which have to fulfil specific functions. Some materials, such as glass, have to be transparent in addition to resisting all weathers. Other materials fulfil such functions as bearing, isolating or accumulating warmth. Because a façade has to fulfil all these functions it is necessary to use more than one material. Even buildings which look as if only one material has been used, for example the glass façade of the offices of Willes, Faber & Dumas Princess in Ipswich, are on closer inspection built up of several materials. In this case sun-preventing hardened glass plates 12 mm thick, rustproof steel patch fittings to attach the glass and silicone sealants between the glass plates.
The necessary presence of various materials in the façade can lead to visual changes which have been observed as a form of soiling. It is for the designer to select adequate materials and give such a form to the façade that minimal façade alterations occur caused by adjacent and additional materials.
In this chapter the façade materials most often employed will be juxtaposed in order to see how the materials can interact under the influence of the external climate.

facade material	bricks	concrete	natural stones	glass	metals	paints	plastics
bricks	....	....	....	....	....	....	....
concrete		....	....	....	....	....	....
natural stone			....	....	....	....	....
glass				....	....	....	....
metals					....	....	....
paints						....	....
plastics							....

NAPIER UNIVERSITY LIBRARY

For the ageing of plastics in buildings use is made of the
tentative recommendations of Rilem TC 34-APB. The
tentative recommendations only cover ageing by climato-
logical agents of the families : PVC, PMMA, polyolefins
(PE and PP), reinforced polyester (UP) and PVC.
The assesment of ageing is based on evaluating
modification in the following properties :

1. physico-chemical properties (P)
2. mechanical properties   (M)
3. properties of appearance   (A)

Physico-chemical properties are composition, mass and
molecular weight. There are also a number of mechanical
properties such as tensile stress, torsion and hardness.
For the scope of façade alterations the properties of
appearance are of more interest. These properties are :

- colour   (c)
- contrast   (g)
- gloss   (b)
- roughness   (r)
- translucence   (t)
- haze   (h)
- soiling   (s)
- decohesion   (d)
- shape stability (f)

Each property has to do with a change in appearance. As
far as is known alteration in plastics is generally caused
more by the influence of the climate than by chemical
attack by products of façade materials. The change in
appearance of plastics as a result of changes in the
properties of colour, contrast, gloss, transparency and
even soiling may be interesting but from the point of view
of façade alterations caused by adjacent and additional
materials, the properties of decohesion and stability of
form are much more interesting. In the different types of
decohesion a distinct classification can be made between :

- homogeneous materials
- composite materials
- laminated materials

Façades can alter through the deposition of products of
the decohesion of plastics and also the deposition of
products of the chalking of paint. Because the
transmission of light through the decohesion products of
plastics is greater than the transmission of light through
the chalk products of paint, the effect of the latter on
the soiling of façades is more important. For this reason
we group plastics and paints together to simplify the
table.

facade material	bricks	concrete	natural stones	glass	metals	paints plastics
bricks	....	....	....	....	....	....
concrete		....	....	....	....	....
natural stone			....	....	....	....
glass				....	....	....
metals					....	....
paints, plastics						....

2.4.2 Material combinations

Bricks - concrete :

The main components of the mortar in the concrete as well
in the mortar layer between the bricks are :
- calcium hydroxide, $Ca(OH)_2$
- alkaline ions K^+ and Na^+
- silicates
- metal ions, e.g. Fe^{++}
Façade alteration as a result of interaction between the
bricks and the concrete will be most affected by calcium
hydroxide and alkaline ions.
Calcium hydroxide :
Hardened portland cement paste contains approximately 30%
calcium hydroxide in the form of small crystallites. The
water in the pores of the hardened cement paste is a saturated solution containing approximately 1.3 g/l calcium-
hydroxide.
   The pH factor of the cement water of approximately 12.5
shows its strong alkalinity. Calcium hydroxide activity
diminishes with advancing age, when the pores in the
hardened cement paste close up.
   Depending on the after-treatment of the concrete and
run-off water as a result of the type of form used,
calcium hydroxide efflorescence of the surface reacts with
carbon dioxide to give the insoluble calcium carbonate,
which causes white blotches and encruststions on the
adjacent material as well as on the material itself.
   Alkaline ions K^+ and Na^+ :
Small quantities of alkaline ions are to be found in the
water contained in the hardened cement-paste pores. They
originate in the materials of the cement manufacture or
possibly in the concrete additives. Also, pigments may
contain alkaline salts that are residues of the
manufacturing process. These salts may cause
efflorescence on the surface of porous building materials
such as bricks. On the other hand, alkaline salts from
the bricks (e.g. Na_2SO_4 and K_2SO_4) can cause staining on
underlying material.

Bricks - natural stone :

Encrustation of calcium hydroxide products on natural
stone occurs if rainfall run-off from the brick is so
great that lime products from the brick mortar are
deposited on the natural stone. Brick itself contains
certain salts which can cause the bricks to efflorescence
but also   discolour the natural stone.

Bricks - glass :

The staining of window glass by run-off water from wall

panels is due primarily to the extraction of alkali silicates from the mortar. These form a deposit of silica on the glass surface, which is hard to remove. The porosity of the mortar plays an important part in this phenomenon.

Beside silica, calcium and alkali-containing materials can be deposited. These stains are easier to remove. Damage to glass by alkali silicates occurs not only in the period shortly after the erection of the building. It can also happen years after the erection when for example renovation of window panes leads to a change in water run-off.

The presence of glass in a façade can have a marked effect on the soiling of the façade. This effect will be strengthened if there are large windows filled with glass panels above brick panels. Even where amounts of rainfall are small such a design will lead to water run-off from glass to brick. For that reason the brick stays longer under wet conditions, thus affecting for instance the growth of algae.

Bricks - metals :

There are three major influences acting between bricks and metals.

First, metals can be altered by the influence of strongly alkaline solutions from the mortar. It is known that alkaline solutions may attack and dissolve aluminium.

Second, metals possess a very dense surface which leads to rapid rainfall run-off and results in clean stripes on brick below vertical metal elements.

Last but not least, oxidizing metal products, if there are any, will be deposited on the porous brick and give rise there to a marked soiling pattern.

Bricks - paints, plastics :

Certain pigments in coatings or dyes may discolour due to akaline reactions of calcium hydrate products from the brick mortar. Moreover, due to saponification of softeners, synthetics may become brittle. Furthermore, the secretion of alkaline salts often causes peeling of paint coatings on building materials.

Paint itself is subject to weathering which can lead to chalking of the paint. These chalk products consist of very fine particles which will be transported by rainfall run-off and deposited on underlying materials. It is very difficult to remove these particles. Architectural vision can lead to a colour for paints which contrasts with the colour of other façade materials, which makes the chalk products more visible.

As mentioned before plastics also cause decohesion products which will give a similar soiling pattern to the

55

chalk products of paint. Normally, however, the decohesion products of plastics are less visible than the paint
products.

Concrete :

The façade alterations caused by concrete on adjacent and
additional materials are quite similar to the effects described for brick mortar.

Natural stone - glass :

Various forms of silicates are to found in calcareous
sandstone. Under certain circumstances these are soluble
in water and segregate on the surface in the form of
coatings on glass which are insoluble in water and acids.

Natural stone - metals :

From the point of view of soiling, natural stone does not
have any influence on metals. If there is any influence
such as an alkaline solution of limestone with aluminium
it has more to do with an agressive attack and destruction
of material than with soiling.
    The metal itself can produce oxidizing products,
depending on the protection and climatical conditions of
the metal. Oxidizing metal products consist of very fine
particles. Rainfall run-off will transport them to each
underlying façade material, including natural stone.

Natural stone - paints, plastics :

Natural stone will be influenced by residues of paints and
plastics as described previously.

Glass - metals :

These materials do not affect each other in a visible way.

Glass - paints, plastics

These materials do not affect each other in a visible way.

    Not all materials which have been and will be used in
façades have been discussed. Nevertheless it may be clear
that designers have to be very careful when using
different types of material, not only from the point of
view of soiling but especially from the weathering
aspects. A clear example of the latter is the use of
different types of metal in a façade which can lead to
electrolysis.
    In conclusion a special material will be mentioned.
This is the group of sealants between for example brick,

concrete and natural stone. Problems can be found
especially in sealants which retain plasticity. To reduce
the stiffness of these sealants mineral oil may be added
to the sealant. It is this mineral oil which migrates in
a porous material and can also be spread by rainfall
run-off to façade material on both sides of the joint. On
these fatty layers dust, soil and soot particles will
stick very well. The same problems occur when using
silicone sealants.

References

Rilem committee 34-APB (1981) Ageing of plastics in buil-
    dings. Materials and Structures 81, 191-227.
de Waal, H. (1981) Staining of window glass by run-off
    water from wall panels. Verres Réfract 35, 305.

2.5 CHANGES BY SOILING

Weathering affects all buildings and all building materials. If men were to return to the moon in ten or even hunderd years much of the delicate scientific equipment left by the Apollo astronauts would still be usable. On the other hand the earth's atmosphere of air and water vapour permits us to live without space suits but destroys delicate equipment if it is left unprotected. Our national environment imposes a life cycle on our buildings and building materials. Slowly or quickly, depending on the material and the characteristics of the atmosphere, all buildings change with time. Weathering then, is the effect of time, the fourth dimension, on our architecture and to ignore it or relegate it to a position of little importance in the design process is foolish.

Perhaps it is not easy to anticipate the succession of youth, maturity and old age through which buildings will inevitably pass but it does seem that many buildings of the last forty or fifty years have been designed with only their youth in mind and have proved quite incapable of gracefully accepting the imprint of the passing years.

This tendency of many modern buildings to look unkempt after a few years is not restricted to specific materials. Often a so-called self-cleansing material will look worse after a few years if it not cleaned than a more robust material in a similar position. The control of weathering involves more than just the choice of the building surface. Design and detailling must combine to control the flow of water on facades or solid parts may need cleaning as often as windows.

There seem to be three basic ways of approaching this aspect of design.

One possible design approach, indicated by a bold broken line in the diagram below, is to design for an eternal youth, defying the attempts of time and the elements to alter the appearance of the building.

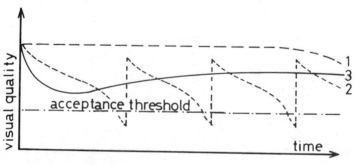

fig. 2.5.1

This might be achieved by the use of expensive materials and very careful design but would never be cheap and would not necessarily be desirable. Is it right for buildings to stand out from their environment for ever - always looking new and never becoming part of the mature and comfortable back-cloth which has been bequeathed to us by former generations?

The second strategy, shown by a lighter broken line, is to design buildings that can be brought back to their original appearance at regular intervals by the injection of a further sum of money. This may mean cleaning or painting or both.

Repainting has been the traditional method for care of many handsome 18th and 19th Century buildings in our towns and cities. It is a useful way of revitalising certain buildings or locations but has two main drawbacks. It commits the building owners to future maintenance expenditure and it presupposes that the building will probably spend a substantial part of its life looking in need of maintenance.

The third option, indicated by a full line, is to attempt to design buildings that can grow old gracefully without expensive maintenance - buildings that will change with time but will not be spoiled. This is probably the most difficult strategy to follow but must be both satisfying and the cheapest in terms of life-time cost.

Many old buildings over the years have developed large accumulations of dirt. In their book 'Salissures de facades', Carrie and Morel illustrate the porch of the church of St.Margaret at Westminster and point out that whilst some people see it as an illustration of the objectionable grime that ravages our cities, others see only the noble patina of age. Certainly a modern structure in a similar dirty state would be likely to receive more critism.

In some cases of course such pollution can harm stone and it must be removed, but where there is no physical damage we can often leave the dirt. Many buildings from former centuries have a visual strength and robustness which is able to carry considerable quantities of dirt without the architecture losing its character. In the more northerly cities of Europe the dirt can sometimes give an emphasis to a design which for much of the year it would lack due to the absence of sharp, welldefined shadows.

And yet closer examination shows that the accumulations of
dirt on facades are not very precisely controlled. By
providing places where dirt can build up without distrac-
ting the eye from the essential characteristics of the
composition the design remains acceptable. By contrast,
many of our more recent buildings are intended as clean
compositions of straight lines and rectangles on which
uncontrolled runs of dirt immediately look out of place.
This is not to recommend the imitation of historic building
forms or the decoration of buildings with superfluous
ornament copied from the past. The illconsidered application of string courses and moulded architraves, without
proper understanding of how they are to control the flow
of water, will do little to help the appearance of a
building. The function that a particular detail has to
perform must first be decided. New details can then be
devised or traditional ones copied from the past, provided
that the chosen detail can be relied upon to perform the
required task.

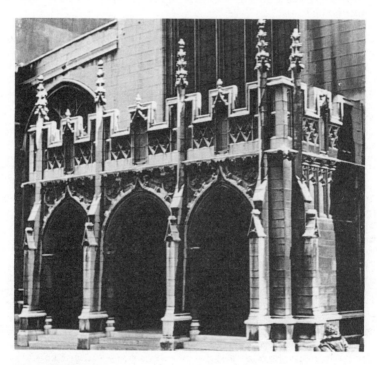

dirt amongst
classical or
gothic details
is less noticeable
than on
modern buildings

An interesting question related to the present subject has to do with the way people react to soiled facades. This question belongs to the field of psychological investigation of architecture, where attempts are being made to analyze the effects of a structure on the behaviour of its users or of passers-by. The ultimate aim is to predict the psychological effects of a given architectural form on people.

With regard to the subject of soiled facades, this type of investigation might concentrate on topics such as the following :

- observation
  how soiled must a facade be before it is noticed by an arbitrary observer?
- cognition
  how much can a facade be soiled before there is a need for information on the origins of the soiling and its technical effects on the facade?
- affective experience value
  what emotions arise from a soiled facade? When is a facade so soiled that it is considered ugly, repulsive and depressive. As a counterpart: is there a point where the soiled facade is so dirty that the characteristic of the total form is considered beautiful, fascinating and fine.
- motivation
  what is the psychological value of a clean facade for the user and the owner of the building?
- behaviour
  what type of behaviour can be caused by a soiled facade ?

Dr. C. Steffen, a psychologist working in the department of Architecture, Delft University of Technology, is conducting research into the following subjects :
    - the experience of soiled facades
    - the experience of colour
    - the experience of texture
In one research project, people's appreciations of the facades of the building of the department of Architecture itself has figured as a subject of research.

south-west facade   north-east facade

In this project subjected experienced was measured with a
technique known as the 'semantic differential'.
    With this technique, subjects are asked to judge the
qualities of an object in terms of a series of seven-point
scales, each having two antonyms as its extreme values.
Antonyms are words with opposite meanings, such as
beautiful vs ugly, or masculine vs feminine. On each
scale, the subject to be judged receives a score between 1
and 7. Scores on the total series of scales provide a
profile of judgements, which is called the index of
experience for the object under consideration. In this
way, it is possible to differentiate between people's
experiences of, for example, clean and dirty facades.
    In the present investigation, a list of 30 antonyms was
used, arranged in such a way that they alternated in intensity : firstly the weak intensity was on the left side
of the scale and the strong intensity on the right side
(e.g., simple .... complex) ;
then the strong intensity on the left side and the weak
one on the right side
(e.g., agitated .... calm).
The reason behind this alternation was to prevent subjects
of scoring automatically and without thinking.
    In this investigation the following qualities were
found to differentiate significantly between people's
subjective experiences of clean and dirty facades,

clean facade	dirty facade
simple	complex
calm	agitated
hard	soft
monotonous	diversified
satisfying	unsatisfying

stimulating	depressive
quiet	busy
artificial	natural
clear	vague
sterile	fruitful
businesslike	playful
tidy	untidy
beautiful	ugly
orderly	confusing
objective	subjective

A factor analysis of these results revealed that subject's experience of clean versus dirty facades was based on six basic variables :
in fact facades were judged in terms of how businesslike, how orderly, how interesting, how complex, how stimulating, and how safe they were felt to be.
For the continued studies a repeated investigation of different buildings should be carried out directed at the difference in experience value of form-identical facades in clean and dirty situations.
Similar investigations could take place with the help of a series of buildings where the facade is clearly distinguished by the intensity of the dirtiness and/or by the different forms in which the dirtiness is found. It is desirable to compare an unlimited soiled building with one where the designer took into consideration the soiled aspect that would arise from the falling rainwater and therefore minimized the soiling aspect.

The atmosphere which as mentioned before makes life possible on this earth, is not simply water vapour and air. It carries a certain amount of solid matter from natural sources - pollen, bacteria and dust swept into the air by the wind or emanating from volcanic eruptions. Unfortunately man's activities have added to this natural pollution great quantities of solid and gaseous products from our fires, industrial processes and traffic. Air pollution is not a new phenomenom but the pattern can be seen to be changing. Complaints about the pollution of London's air, for instance, began with the introduction of coal for lime burning and smelting in the thirteenth century. When coal started to replace wood in domestic hearths in the seventeenth century air pollution rapidly reached serious levels.

Although increases in pollution have only recently been measured, its effects on buildings and vegetation and on the light and weather of cities have been recorded in literature and paintings for centuries. Throughout the industrial revolution the problems grew. Within the last hundred years there has been a slight amelioration as a result of improved transportation allowing a sudden expansion and consequent lessening in density of the build-up areas of cities. In the last twenty-five years legislation has produced dramatic improvements in the levels of air pollution in urban areas.

All European countries now have some clear air legislation but severity of standards and strictness of enforcement differ widely. Most countries aim to control the emission of pollutants at known sources such as factory chimneys and vehicle exhausts but the European Community intends to establish standards of air quality for urban areas irrespective of sources. This will entail monitoring levels and presumably publishing information that will be of use archtects attempting to predict the weathering of buildings.

Chapter Three

Non-biological soiling

3.1 AIRBORNE PARTICULATE MATTER AND ITS MEASUREMENT RELEVANT TO THE SOILING OF FACADES

3.1.1 Introduction

Soiling of façades is that complex of processes which takes place at the surface of façades and leads to a change in the general appearance of the façade. This is undesirable as a rule and sometimes expensive measures of prevention or cleaning have to be undertaken. Soiling can be either a discoloration of regions of façades due to the deposition of airborne material such as soot, or it can be attack of the building material with subsequent localized losses of material. In all cases, soiling is a disfiguration which can reach a serious extent after 10 years' exposure to the atmosphere. An interested observer, however, would already observe soiling patterns after a year or so (1).

Altogether, it is a process of very inefficient deposition of air pollutants accumulating at the façade surface over many years.

Soiling of façades by exposure to pollutants is the result of a sequence of occurrences starting with emission of the pollutant by some source, followed by atmospheric transference and, ultimately, deposition onto the façade surface.

This chapter will focus on the deposition and related processes, the resulting effect of soiling and the relevant measurement of the pollutant causing this effect.

However, it is convenient to give some attention to other aspects of the problem, not least because everything is related to everything.

3.1.2 Air pollution

3.1.2.1 Aerosols

Ambient air to which building façades are exposed, consists of pure air (defined e.g. by (2)) contaminated by various polluting gases and particulate compounds. Gaseous pollutants include SO_2, NO_x, H_2S, NH_3 and O_3. Particulate matter is even more multicomponental :

sulphates, nitrates, soot, silicates, metal cations, organic compounds, etc.. Table 3.1 gives an impression of the composition of atmospheric aerosols. Trace constituents are presumably of less importance for soiling. Fig. 3.6 gives further information about the contents of bulk components in ambient aerosols ; this will be treated in more detail below.

Depending on the regional and local climatology the contaminants are transported to the façade surface either by dry deposition or by wet deposition (in precipitation).

The velocity v of dry deposition of a pollutant is defined as the ratio of the pollutant flux deposited onto a surface to the pollutant concentration at a certain distance from that surface. Reactive gases such as SO_2 and NO_2 have v_d - values of the order of 0.1 up to a few $cm.s^{-1}$ Particles have v_d-values of similar magnitude in the extreme fine and extreme coarse size range (fig.3.1.2.1. 1). Particles in the intermediate size range of 0.1 µm to a few µm possess extremely low deposition velocities (10^{-2} - 10^{-3} $cm.s^{-1}$). Since residence time and transport distance are inversely related to v_d, this intermediate size range has extremely large residence time (several months) and long transport distances (ca. 10.000 km). Fig. 3.1.2.1. 3 illustrates this for various particle size categories and gas reactivities.

Transference of pollutants to the Earth's surface can also be due to wet deposition. Incorporation into the depositing wet phase can be either in-cloud ("rain-out") or below-cloud ("wash-out"). Fig. 3.1.2.1. 2 represents the size dependence of incorporation in rain droplets. Note the striking similarity between Figs. 3.1.2.1 1 and 3.1.2.1 2 which is due to the identical size-dependent processes involved. Coarse particles are deposited or incorporated due to their inertia ; the fines, however, diffuse (by Brownian motion) to surfaces and droplets. Table 3.2 gives a schematic overview of the various removal processes for atmospheric aerosols. As a result, due to the close relation (high diffusion coefficient, difficult resorption) between reactive gases and the very fine end of the particles, it will be clear that those gases are removed mainly by dry deposition*.

According to Whitby (3) the size spectrum of atmospheric aerosols is composed of three modes : transient nuclei, accumulation range, and coarse particles mode (Fig. 3.1.2.1. 4). The two extreme modes (< 0.1 µm and > 2 µm) are related to the two principal modes for particle formation : molecular and mechanical processes, respectively. The intermediate mode results mainly from coagulation of transient nuclei particles. From the above

*Excellent reviews of dry deposition and wet deposition of gases and particles from the atmosphere are given in (23) and (24) respectively.

description of particle removal from the atmosphere, it is obvious that the latter mode is the most persistent one, and is mainly removed by wet deposition.

In view of the effects of aerosols, volume (or mass) /size distributions are more relevant because the composition of the various size fractions differs greatly as a result of the specific contribution of various aerosol sources to the various particle size fractions. The mass/size distribution of the atmospheric aerosol is usually bimodal with two modes : from about 0.1 µm to several µm (the accumulation mode, see above) and beyond several µm (the coarse particles). The mass in the coarse particle mode can be dominant in the atmospheric aerosol mass, namely under conditions of dry, windy weather (fig. 3.1.2.1. 5). The accumulation mode is largely composed of sulphates, nitrates, ammonia, soot and organic material (see fig. 3.1.2.1. 6). Coarse particles resemble soil composition (Ca, Si, Fe, Al, etc.) to a large extent. In coastal areas seasalt particles (Na and Cl) can contribute considerably to the coarse particle mode and to a lesser extent to the submicron regime.

Of major importance for façade soiling is the presence of light-absorbing materials, mainly soot, predominantly in the submicron size range and coal dust and iron ore in the coarse particle mode. These coarse particles have only short residence times in the atmosphere and their influence is therefore rather local. Soiling of façades due to deposits of coarse particulate material is only of importance in the direct surroundings of their sources, namely near stock piles and terminals handling those powders.

A good illustration of the general presence of light-absorbing particles in the accumulation mode only, can be obtained from comparing the filter samples of so-called dichotomous samplers. Such samplers collect particulate material separately in two particle size fractions with a cut-off point near 2 µm diameter. Fig. 3.1.2.1. 6B shows the grey-coloured filter loading of the < 2.5 µm fraction, the coarse fraction being white coloured. This colour difference is a general observation in rural areas and also in urban/industrial areas (however, not in the vicinity of stock piles etc.).

The optical properties of atmospheric particles show an overwhelming contribution from the 0.1 micron - 1 micron size fraction. This is due to the fact that the surface to volume ratio of particles decreases with increasing particle size whereas below a few tenths of a micron the light interaction drops sharply (Rayleigh scattering). The very fine soot particles are, therefore, optically much more active than the coarse coal dust particles.

From extensive studies of the composition and optical properties of atmospheric aerosols, it is evident that soot causes the black colouring of the submicron fraction

(4). Also, the soiling of façades is the result of this
ubiquitous soot-containing atmospheric aerosol. Schaffers
(5) reports that soot fills the surface pores of many
sandstones ; Spence and Haynie (6) have shown that soot
deposit is responsible for the blackening of painted surfaces at various U.S. locations. There is a close relation between the atmosphere's soot content and façade
soiling.

3.1.2.2 Soot

An excellent review on atmospheric soot can be found in
the 1980 GM symposium on Particulate Carbon (9).
    Soot in the atmosphere is a by-product of incomplete
combustion.   It is present in the submicron size range
and, hence, has a long atmospheric residence time and
considerable amounts of soot are present (9).
    From the importance of soot for the (optical) effect of
soiling one may correctly infer an important role for soot
in atmospheric optics, namely in visibility.   Table 3.3
shows that light absorption by soot contributes
significantly to light extinction in the urban atmosphere.
Soot particles are aggregates of primary particles of
graphitic material.   The automobile aerosol, which is soot
to a large extent (9), consists of aggregates of several
hundred primaries of 0.05 µm geometric mean diameter (10).
    The specific light absorptivity of soot is 10 $m^2.g^{-1}$
(11).   From this, combined with an atmospheric average
soot concentration of 0.1 - several $µg.m^{-3}$ (11, 12) and an
average dry deposition velocity of 10^{-2} $cm.s^{-1}$, an average
yearly surface obscuration by soot of about 10 %.yr^{-1} can be
calculated.   This is quite reasonable in view of the rate
at which the soiling effect arises.
    The general observation that also downward facing sur-
faces are blackish, nicely fits the aerodynamic properties
of soot-containing ambient particles : these are in the
size range 0.1 µm to a few µm having omnidirectional
deposition characteristics.   The soiling patterns of
façades are the result of the eroding action of rainfall
run-off interacting locally with the uniform soot deposit
(1).
    As would be expected, according to the efficiencies of
the various atmospheric removal processes (Table 3.2), the
submicron soot particles are deposited mainly in precipitation.   Soot concentrations in wet deposition are about
200 µg per litre (18).   This deposition of soot onto
façades, however, does not lead to important soiling,
taking into account the observed uniform deposits of soot.
It is probable that the attachment of soot particles to
the façade surface has a low efficiency under wet conditions.
    There are no in situ soot monitors although some are in
a promising stage of development (13). Automatic disconti-
nuous monitoring of soot using light absorption of filter

loading is realized in the so-called aethalometer (12) ;
the present time resolution is 5 minutes. A widely used
measuring technique is the "black smoke" method : filtra-
tion and measurement of the reflectivity of the filter
loading. This is an old method ; it is standardized in
the UK (14), in OECD countries (15), by ASTM (16) and is
under development in ISO. These reflectance data of filter
papers (darkness index) are a proper measure of the soot
content (17). As soiling is a long-term effect (of the
order of a year or more) there is no need for real-time
soot measurement with a high time-resolution. Perhaps the
good old "black smoke" method fits these requirements
sufficiently well.

3.1.2.3 Coarse particulate matter
As argued above, coarse particles can lead locally to
serious façade soiling. Well-known examples are coal dust
and ore dust mainly from coal and-ore handling facilities.
   The main problems of measurement of such coarse particulate matter are sampling them correctly. Moreover, representative sampling is not adequate in view of the
strong dependence of coarse particle deposition on meteorology. Sampling tehniques are required here which are
more relevant to deposition than extractive sampling, even
when this is achieved representatively.
   One sampling method should be mentioned which is principally correct for soiling assessment, namely deposition
techniques such as dustfall gauges. The method involves
exposing a bucket for a period of about one month and
analysing the contents. There are various designs which
are standarized in various countries (22).
   One aspect of the dustfall gauges is that they collect
dry and wet deposition. The latter is not desirable, probably because particulate matter transported to façade
surfaces is not attached too efficiently to those surfaces
under wet conditions.

3.1.2.4 Other pollutants
So far we have not discussed the part played in façade
soiling of various other (gaseous) air pollutants.
   The contribution via different mechanisms from
pollutants such as SO_2, NO_x, NH_3 and rain constituents is a
separate subject which will not be dealt with extensively
here. Nevertheless, it should be emphasized that there is
an important interaction between soot and gypsum (reaction
product of building material with SO_2 and acidic sulphates)
on the façade surface (1, 7, 8).
   The most reactive species of atmospheric pollution are
SO_2, NO_2, H_2SO_4, HNO_3, NH_3, ammonium nitrate and ammonium
sulphates. Of these, H_2SO_4 and the ammonium salts are particulate and have submicron sizes. As a result, their deposition velocity is low compared with reactive gaseous

69

pollutants, which combined with the similar order of magnitude of concentration levels makes it reasonable to infer that particulate deposits will not contribute importantly to secondary soiling (chemical attack). Of the above given list of gaseous pollutant species HNO_3 is the most reactive, being present at levels of 1-30 $\mu g.m^3$.

Nitric acid vapour will react rapidly with the façade surface. However, the alkalinity of this layer will probably remain sufficiently large not to produce an important bias to SO_2 uptake. The oxidation product of SO_2 adsorbed at the façade surface, namely sulphate, is found extensively. The solubility in precipitation of these reaction products from dry deposition is a very important aspect of façade soiling and weathering.

H_2S, though relatively unreactive, may react with certain metals (e.g. Fe) in the façade surface and may lead to dark-coloured reaction products. Even superficial reaction with iron-containing particles in the façade surface can yield significant soiling. This is an effect unexplored hitherto.

3.1.3 Pollutant measurement

An excellent survey of monitors for pollutants is to be found in (21). Measurement of air pollutants in relation to façade soiling can be subject to a wide variety of errors and mistakes of principle.

a) The most adequate measurement of façade soiling is measurement of the soiling of the façade itself by principles which are relevant to the effect. This is the only solution when secondary soiling also has to be taken into account, because of its complex physicochemical nature. A still somewhat abstract approach could be the exposure of specimens of façade-like materials. These should be placed in the direct vicinity of the building concerned. Various geometries and shapes have to be used in order to eliminate the relation between deposition velocity and spatial dimensions of the deposit specimen. In principle, this is the approach proposed by Theissing (19). Further analyses of the samples obtained in this manner could comprise : reflectance measurement, quantitative chemical surface analyses, and microscopic analyses. In those cases where secondary soiling is of minor importance and also where hypermicron soiling particles are absent, the situation is easier. Under such conditions measurement of the airborne (soot) concentration could suffice. Combination of this concentration with an averaged deposition velocity of about $10^{-2}cm.s.^{-1}$ yields adequate soiling information.

b) Sampling can be inadequate. When the soiling
    pollutant consists of coarse particulate (> 5 μm)
    matter, so-called anisokinetic sampling
    (aspiration) can lead to large errors. Anisokine-
    tic sampling strictly is due to particle inertia
    when particles are unable to follow the streamlines
    in the case of a sampling velocity different from
    the general velocity of the particle movement.
    Anisokinetic sampling can also occur when sampling
    under a non-zero angle with respect to the particle
    movement. Fig. 3.1.3 1. gives an impression of
    various effects and situations. Sampling errors
    may also arise in the case of sampling line losses,
    especially for too long sampling ducts.
    As indicated above, soot will show no sampling
    errors because of its small particle size (0.1 -
    1 μm).
c) Sample analysis can be inadequate. The main
    problem here is to find a useful specific analy-
    tical technique. The small sizes of the samples of
    ambient particles and the ease of some non-specific
    analytical techniques (such as weighing) have al-
    ready led generally to undesirable practices such
    as health assessment of air quality by sample
    weighing ("total mass" is irrelevant hygienically).
    Because soiling is an optical effect we advocate
    sample analysis by optical principles : light
    absorption or reflection. In the above mentioned
    case of soot analysis we have already listed the
    various current techniques.

Finally, we comment on the inadequateness of the so-called
High-Volume Sampling (HVS) for Total Suspended
Particulates (TSP) which stands for sampling about 2000 m
in 24 hours and weighing the sample. TSP is irrelevant to
soiling because :

  - it involves sampling of coarse particulate matter
    which is meteorology-dependent in a manner which
    differs in principle from soiling ;
  - weighing is not an adequate analytical technique for
    characterizing air quality in view of the soiling
    potential of ambient air.

3.1.4 Summary and conclusions

3.1.4.1
Soiling of façades by exposure   to pollutants is the result of a sequence of occurrences starting with emission
of the pollutant followed by atmospheric transference and,
finally, deposition onto the façade surface. The exposi-
tion flux is the product of the pollutant's concentration
near the façade and its deposition velocity onto the
façade. Both are complicated functions of meteorological
conditions, for example.

3.1.4.2
Two kinds of soiling mechanism must be distinguished :
- primary (physical) soiling due to deposition of
  mainly particulate matter of a colour different from
  that of the façade surface ;
- secondary soiling due to deposits which interact with
  the façade surface or are influenced by micromete-
  orology.

Secondary soiling is an extremely complex physicochemical
effect which at the moment can only be assessed adequately
by in-situ determination. Primary soiling, however, being
a purely optical effect, could in principal be monitored
in the surrounding atmosphere by measuring the chemical
species responsible for this effect. Soot and fugitive
dust emissions (coal and ores) are the dominant soiling
pollutants.

3.1.4.3
Atmospheric aerosols consist of two size fractions with a
saddle point at about 2 µm. The submicron fraction is of
molecular origin (mainly fossil fuel burning) and has a
long atmospheric residence time. The hypermicron fraction
is mainly wind-blown and is transported only locally.

3.1.4.4
Primary soiling is the result of dry deposition of soot
which is ubiquitous. From atmospheric optics, soot
concentration data and an average dry deposition velocity
for soot, an average soiling rate of about 10 % per year
can be calculated. Wet deposition of soot is of
negligible importance for façade soiling. Soot monitoring
can be done with the well-known "black smoke" method
(filter sampling and reflectance measurement).

3.1.4.5
Important soiling due to coarse particles (coal dust, iron
ore dust, etc.) can occur locally. Adequate monitoring is
impossible nowadays. Simple and qualitatively correct is
the use of horizontal dustfall gauges (exposing buckets

and analysing the contents afterwards).

3.1.4.6
Various measuring errors have to be considered in air pollution monitoring. Sampling (anisokinetics and sampling line losses) is a problem in the case of hypermicron particles. Sample analysis (non-specific analysis) can introduce large faults. Most adequate measurement is by analysis of the façade surface soiling itself using optical techniques. Usually, this is inconvenient. The second best approach is utilizing standarized specimens of façade material exposed in a standarized way in the vicinity of the façades.

References

(1) Beyer, Oscar, (1980) Weathering on External Walls of concrete, CBI-11:80.
(2) Handbook of chemistry and physics, (1974/1975) Chem. Rubber Cy Press, 55th edition, p. F195
(3) Whitby, K.T. and Cantrell, B. (1975) Atmospheric Aerosols-Characteristics and Measurements, Proc. Int. Conf. Envir.Sensing and Assessment, Las Vegas (Nev.) (IEEE, 1976).
(4) Wolff, G.I. (1981) J. Air Poll.Control Ass., 31,935.
(5) Shaffer, R.J. (1932) The weathering of natural building stones, London HMSO.
(6) Spence, J.W. and Haynie, F.H. (1972) Paint Technology and Air pollution : A survey and economic assessment, EPA.
(7) Gauri, K.L. Proc. Conf. Polluted Rain (Ed. T.Y. Toribara, etal), Univ. Rochester, Rochester (Plenum Press New York).
(8) Gauri, K.L. and Holdren Jr. G.C. (1981) Envir. Sci. Technol., 15, 386.
(9) Particulate Carbon : Atmospheric Life Cycle, (1982), Proc. G.M. Symp. (Eds. G.T. Wolff and R.L.Klinisch), Plenum Press, New York.
(10) van de Vate, J.F. (1977) etal. Proc. IUAPPA Fourth Int. Clean Air Congress, May 16-20, Tokyo, Japan.
(11) Wolff, G.T. (1981) J. Air Poll. Control Ass.,31, 935.
(12) Hansen, A.D.A. (1982) etal.Applied Optics, 21, 3060.
(13) Truex, T.J. and Anderson, J.E. (1979) Atmosph. Envir., 13, 507.
(14) British Standard 1747 (Part 2), (1969) Methods for the Measurement of Air Pollution.
(15) OECD-report Methods of Measuring Air Pollution, Chapter Two (Smoke), Paris, France (1964).
(16) ASTM D1704-61 (1969) Particulate Matter in the Atmosphere (Optical Density of Filtered Deposit).
(17) Brosset, C.A. (1983) private communication (IVL, Göteborg, Sweden).

(18) Ogren, J.A. Deposition of Particulate Elemental Carbon from the Atmosphere ; in (9).

(19) Theissing, E.M. (1983) private communications, Central Laboratory Bredero.

(20) Air Quality Criteria for Particulate Matter, U.S. Dept. of Health, Education and Welfare, (1969) AP-49.

(21) Air Sampling Instruments for Evaluation of Atmospheric Contaminants, 5th Ed. (1978) Amer.Conf. Governm. Industrial Hygienists, P.O.Box 1937, Cincinnati, OH 45201, USA.

(22) E.g.
in USA : Dust Fall Jar and Stand, ASTM D1739 ;
in FRG : Berghoff and Hibernia Gauges, VDI 2119-1 and 2119-2;
in France : Norme francaise NF-X43-006;
in UK : British Standard BSI-1747.

(23) Sehmel, G.A. (1980) Atm. Envir., 14, 983.

(24) Pruppacher, H.P. and Klett, J.D. (1978) Microphysics of Clouds and Precipiation (Reidel, Dordrecht,Holland)

(25) Slinn, W.G.N. (1977) Water, Air and Soil Poll. 7, 513.

(26) Husar, R.B. and Patterson, D.E.(1980) Synoptic Scale Distribution of Manmade Aerosols, in Proc. WMO-Symp. Long Range Transport of Pollutants, etc. (Sofia, Bulgaria, 1-5 Oct. 1979) WMO-538.

(27) Patterson, E.M. and Gillette, D.A. (1977) J. Geophys. Res., 82, 2074.

(28) Mészaros, A. (1977) Atmosph. Envir., 11, 1075.

(29) D'Almeida, G.A. and Schütz, L. (1983) J. Climate Appl. Meteorol., 22, 233.

(30) Shaw, R.W. and Stephens, R.K. (1980) Trace Element Abundances in Aerosols : Anthropogenic and Natural, Sources and Transport (Ed. Kneipp, T.J. and Lioy, P.J.) Ann. New York Ac. Sci., 338, 13.

(31) Belyaev, S.P. and Levin, L.M. (1974) J. Aerosol Sci., 5, 325.

(32) Durham, M.D. and Lundgren, D.A. (1980) J. Aerosol Sci., 11, 179.

(33) Stöber, W. (1979) Sampling and Evaluation of Aerosols under Biomedical Aspects, Proc. G.A.F. Symposium, Düsseldorf, FRG.

Table 3.1. Arithmetic mean and maximum urban concentrations of pollutants in air in the United States, biweekly samplings, 1960-1965 (20).

Pollutant	Concentration ($\mu g/m^3$)	
	Arith. average	Maximum
Suspended particulates	105	1254
Fractions :		
Benzene-soluble organics	6.8	
NO	2.6	39.7
SO	10.6	101.2
NH	1.3	75.5
Sb	0.001	0.160
Sb	0.001	0.160
As	0.02	
Be	< 0.0005	0.010
Bi	< 0.0005	0.064
Cd	0.002	0.420
Cr	0.015	0.330
Co	< 0.0005	0.060
Cu	0.09	10.00
Fe	1.58	22.00
Pb	0.79	8.60
Mn	0.10	9.98
Mo	< 0.005	0.78
Ni	0.034	0.400
Sn	0.02	0.50
Ti	0.04	1.10
V	0.05	2.200
Zn	0.67	58.00

Table 3.2 Removal of particles from atmospheric aerosols : removal efficiencies (+ = important ; − = unimportant) of various deposition processes for various size fractions of atmospheric particles.

Deposition proces		< 0.1 μm (incl. reactive gases)	0.1μm<d<2μm	> 2μm
rain-	nucleation	±	+	+
out*	coagulation	+	±	+
wash-out**		+	−	+
dry deposition		+	−	+

*in-cloud removal
**below-cloud removal

Table 3.3 Fraction of light extinction of the atmosphere due to absorption (11)

urban air	0.35 − 0.50
rural air	0.13 − 0.27
remote areas	0.05 − 0.11
Arctic air (spring)	0.4

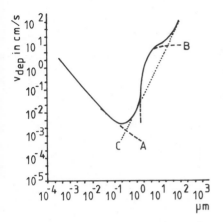

fig. 3.1.2.1. 1
Deposition velocity (v_{dep}) under dry conditions of gases
and particles. v_{dep} of gases is largest for reactive
gases. Deposition velocities are dependent on the
receptor surface roughness and wind velocity ;
particularly, impaction (and interception) has a
meteorology dependent efficiency. A = diffusion, B =
impaction (interception) ; C = gravitational deposition
(interception). Diffusion dominates dry deposition of
reactive gases (SO_2, O_3, etc.) and of < 0.1 µm particles.

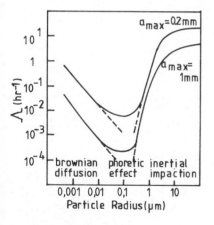

fig 3.1.2.1. 2
Scavenging coefficient (Λ, removal of particles by rain
drops) as a function of particle radius and drop size
(a_{max}).
Note the similarity between figs. 1 and 2 which is due to
mechanistic analogies (diffusion and impaction).

77

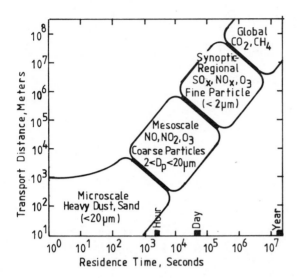

fig. 3.1.2.1. 3
Residence time and transport distance of various air constituents. Due to their lower deposition velocities of
less reactive gases (CO , CH) and fine particles, their
residence times, etc. are larger.

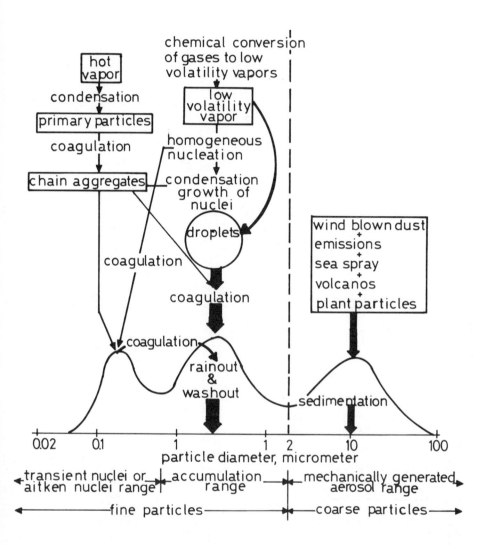

fig. 3.1.2.1. 4.
Schematic size distribution of atmospheric aerosols,
terminology of size fractions, and principal processes of
formation and removal. Note the different origins of the
size fractions > 2 μm and < 2 μm : coarse particles are
mainly terrestrial and mechanically generated, fines stem
from molecular processes and are mostly anthropogenic.

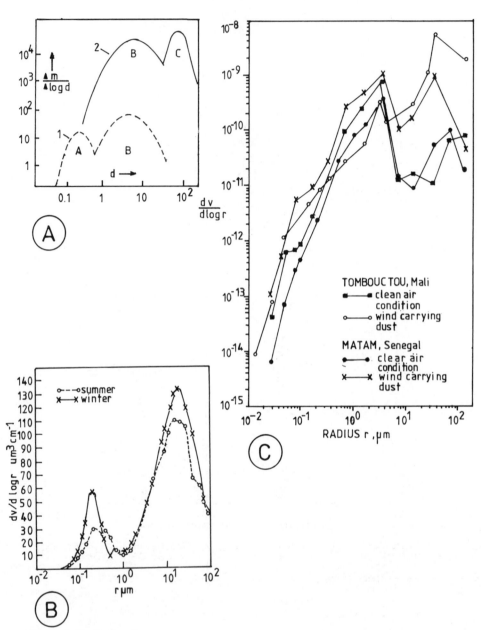

fig. 3.1.2.1. 5
Typical mass (or volume) size distributions reflecting
meteorology dependence. A shows the shift to larger sizes
at higher mass concentrations which result under dry and
windy conditions ; B : Budapest, Hungary ; C : desert
aerosol (windy and dry conditions - dustwind - lead to
increased coarse particle mode).

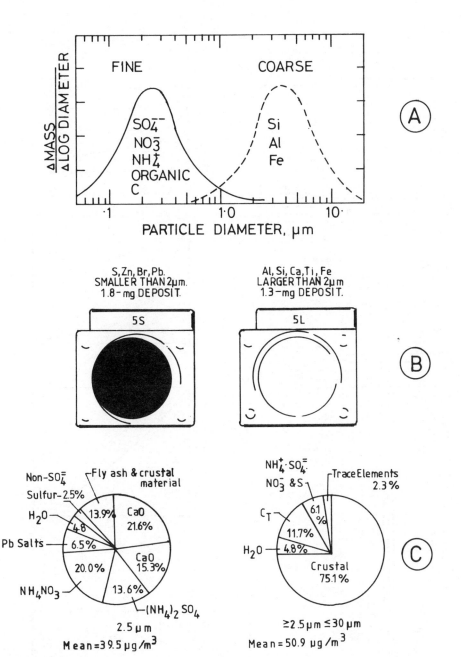

fig. 3.1.2.1. 6
Composition of the bimodal size distribution of the
atmospheric aerosol (A and C). B shows the dark appearance
of the below 2 μm size fraction which is due to the
presence of soot in the filter loading. Soot is also
responsible for façade soiling.

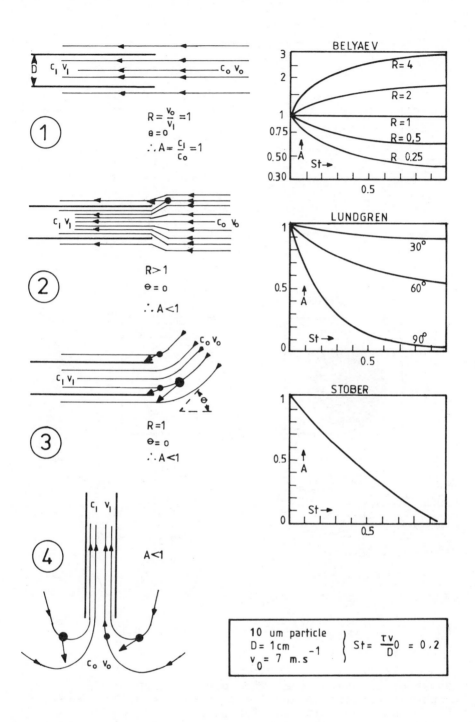

fig. 3.1.3. 1.

3.2 TRANSFERENCE OF ATMOSPHERIC POLLUTION TO THE FACADE

3.2.1 Introduction

Numerous phenomena influence the movement of pollutants in the atmosphere over varying distances.

The description of these phenomena in this chapter especially concerns all solid pollutants whose behaviour is similar to that of gases contained in smoke. This implies that the diameter of these particles is below approximately 20 μm, according to the references. Larger particles form a deposit near the emitting source.

This chapter describes the meteorological factors which influence the movement of pollutants. Two examples are given from a more experimental point of view and the principal means for laboratory study are examined.

3.2.2 Influence of meteorological conditions

The transference in the atmosphere of pollutants given off by one or several sources is directly influenced by atmospheric weather conditions such as wind, temperature, state of the atmosphere and atmospheric water (rain, fog).

3.2.2.1 Wind
Wind is caused by the displacement of air from high pressure zones to zones of lower pressure. Wind speed and force vary according to this difference of pressure and friction forces (roughness of the site).

The characteristics of the wind have an effect on the diffusion of pollutants. In an article on atmospheric pollution (cf references), T. Bourelly of the Centre Scientifique et Technique du Bâtiment, Paris, suggested that pollution is concentrated in a small area if wind speed is low, whereas strong winds will "dilute" pollution, making it less concentrated but spread over a greater area.

3.2.2.2 Temperature of the atmosphere
As a first approximation, the atmosphere is considered to be a thermodynamic system, which means specifically that a particle which rises tends to cool, since the pressure decreases and inversely ; the process is said to be adiabatic.

Under these conditions the atmosphere is indifferent and the displacement of the particle studied is linked to the initial conditions of displacement (fig. 3.2.2.2. la in the appendix).

In this case, a particle which is rising remains warmer than the atmosphere and continues its movement upwards.

The atmosphere is unstable when the conditions are sur-

83

adiabatic, in other words, the decrease in temperature as the particles rises is greater than the dry adiabatic gradient (fig. 3.2.2.2. 1B and 1C in the appendix).

The atmosphere is stable if the vertical decrease in temperature is less than the dry adiabatic gradient. For these conditions, any movement given to the particle tends to bring it back to its initial position (fig. 3.2.2.2. 1D in the appendix).

3.2.2.3 Temperature reversals

When referring to the three previous states of the atmosphere, we have considered a regular decrease in temperature.

However, it can happen that the atmosphere becomes warm at altitude, the temperature being higher than in the lower strata. Such conditions are met with in very cloudy or foggy weather when the sun warms the higher strata more easily than those near the ground, which are thereby less accesible to solar radiation. This leads to temperature reversal (fig. 3.2.2.3. 1 in the appendix).

As the sun warms the atmosphere and the fog dissipates, the reversal cancels out. This is observed a few hours after dawn. Nevertheless, in winter, temperature reversal can persist for several days.

Temperature reversal has a decisive effect on the incidence of pollution, because on the one hand the atmosphere is stable under the reversed ceiling and on the other the phenomenon of the natural ascension of polluting particles is slowed down. These conditions lead to an increase in pollutant concentration.

This reversal phenomenon can be frequent, for example more than 200 reversals are registered in Paris over one year ; some of these can last several days.

3.2.2.4 Atmospheric water

Generally speaking, evaporation and condensation phenomena have the effect of picking up atmospheric pollutants. In addition, the particles coagulate and their concentration increases in foggy weather.

In contrast, rain cleans the atmosphere by precipitating the pollutants to the ground. According to Detrie (cf. references) steady rain falling at a rate of 1 mm/h for 15 minutes will eliminate 28% of the particles measuring about 10 μm.

This effectiveness decreases with the size of the particles, the limit being approximately 2 μm.

3.2.2.5 Explanatory figures

To make the explanation quite clear, we have annexed figures showing the effect of the incidence of meteorological conditions on the diffusion of atmospheric pollutants. These illustrations are taken from Detrie and Facy (references).

3.2.3 Diffusion of atmospheric pollutants given off by a chimney

Studies related to industrial-type chimneys have produced numerous formulae for calculating the height in relation to meteorological parameters, the intrinsic characteristics of the emission and of the surelevation of fictive height (ascension of the smoke plume).

The pollutants present in smoke are essentially gas and dust. It can be assumed that particles of diameter below 20 μm behave similarly to the gases surrounding them, whereas the larger particles are subjected to gravity and quickly separate from the plume, becoming deposited on the ground at short distances.

The example presented below is an expression of the Sutton formula. It offers only one approach to the problem through the simplified hypotheses proposed. Moreover, most of the formula used for these calculations bring in different coefficients to be determined experimentally and which take into account the above mentioned climatic parameters, the relief etc.

The problem raised is therefore to know for a chimney height h at what distance the pollutants (dust) will be deposited.

Amongst the simplifying hypotheses for dealing with this example, we consider a flat and homogeneous site, a steady wind (4-5 m/s), normal turbulence, low temperature gradient and stable atmosphere.

The Sutton formula can be applied with these restrictive conditions :

$$C = \frac{Q}{\pi C_y C_z u x^{2-n}} \cdot e^{-\frac{y^2}{C_y^2 x^{2-n}}} \left[e^{-\frac{(z-h)^2}{C_z^2 x^{2-n}}} + e^{-\frac{(z+h)^2}{C_z^2 x^{2-n}}} \right]$$

$$C = \frac{2Q}{\pi C_y C_z u x^{2-n}} \cdot e^{-\frac{1}{x^{2-n}} \left(\frac{y^2}{C_y^2} + \frac{z^2}{C_z^2} \right)}$$

height of chimney equal to 0

85

C   : concentration of dust at a distance x from the
        origin
Q   : emission rate, mass of dust given off per unit of
        time
h   : height of the chimney
n   : dimensionless constant
Cy,Cz: coefficients of lateral and vertical diffusion
u   : wind velocity

The calculation for 50 to 100 m high chimneys is
undertaken as follows :

$$Cy = Cz = 0.10$$

The maximum concentration on the ground is :

$$Cmax = \frac{QC_z}{e\pi h^2 u C_y}$$

and the distance :

$$D_x = \left(\frac{h^2}{C_z^2}\right)^{\frac{1}{2-n}}$$

For a mean value of n = 0.25, the maximum concentration of
dust on the ground is shown to be obtained at a distance
about 20-30 times the height of the chimney.

These formulae are only approximate   as are the field
observations, but they point to certain conclusions :
- the concentrations of dust on the ground are in proportion to the rate of emission and independent of
  the concentration,
- the concentration on the ground is inversely proportional to the wind speed,
- the maximum concentration is inversely proportional
  to the square of the height of the chimney and the
  distance is proportional to h^2,
- the concentration far distant from the source is inversely proportional to the square of the distance
  from the source.

3.2.4 Study of wind diffusion of pollutants given off by
      an extended source

This example is part of a study undertaken by the C.S.T.B.
(Centre Scientifique et Technique du Bâtiment) on
pollution given off by the urban centre of Paris.   It is
interesting since it illustrates the case of an extended
source : 20 km diameter.

With wind from a constant direction, the pollution
spreads in a 20 km wide trail (fig. 3.2.4.1).

Figure 3.2.4.1 : Pollution given off by the Paris
built-up area, likened to a vertical
cylinder of diameter 20 km. Flow in a
20 km wide trail defined by the planes tangent to the cylinder and
parallel to the direction of the wind.

Let us consider a point L located in this trail
(fig. 3.2.4.2). The pollution will be all the higher as
the point L is located near the source (circle with centre
P).
  Experiments have shown that over a year the direction
is not constant and the winds change.
  The angle indicates the maximum variation in the
direction of wind so that L remains in the pollution
trail.
  It is shown that pollution L is all the more intense as
α is wider (L near P). In contrast, for changing winds,
L is all the less exposed as it is further distant from
Paris (small α). Consequently, the determination of the
theoretical degree of pollution must take into account the
frequency of winds from the sector involved and the
proximity of the source.

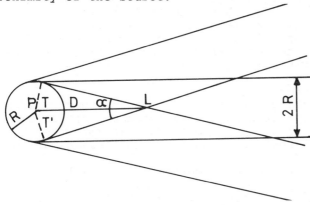

Figure 3.2.4.2 Exposure of a point L to wind-borne
pollution coming from Paris.

Field measurements confirm these hypotheses. For example, on 15 February 1967 with a north-easterly wind and a sector of 60°, the urban centres located in these trails were subjected to about 5 times more atmospheric pollutants than the yearly mean.

Other experiments have been undertaken, as for example in the Laco region (south-west France), and pollution maps drawn up from the findings.

3.2.5 Study of pollutant diffusion by simulation

3.2.5.1 Numerical models
To evaluate pollution diffusion in an urban area for example, each source is considered separately. In addition, coefficients take into account the meteorological conditions forecast in the area under study (speed of wind, temperature).

By introducing other corrective factors, it is possible to apply these equations to different types of relief and environment (city, forests.....).

The study of pollutant diffusion using models implies total respect of similitude particularly as regards the geometric scale and physical parameters.

Experience shows that it is possible to reproduce in a satisfactory way in the wind tunnel (or hydraulic jet) certain characteristics of the atmospheric airstream over urban areas.

Another interesting application would be to study the influence of main road traffic on the production and diffusion of pollutants. The different parameters already described : wind (direction, speed), state of the atmosphere, temperature reversal and volume of traffic are to be considered in these calculations.

Moreover, a main road is the equivalent of a wide source. The pollutants given off are essentially gases. However, small drops of oil (breather pipes, fog) and dust of mineral origin detached from the pavement or vehicles are not to be excluded.

3.2.5.2 Wind models
Most studies on pollution diffusion require that numerous parameters be taken into account : turbulent diffusion, orography, grouping of buildings. To deal with these studies which are sometimes complex, it is essential to set up models.

The equations may become complicated and call for computer processing. However, it is useful to adopt numerical methods to determine the effect of the incidence of different projects on pollution, for example, extension of existing urban or industrial centres.

The limitation of these models is essentially linked to the simplifying hypotheses concerning the different physical parameters.

Numerous investigations have been carried out in France by the Météorologie Nationale (national meteorological office), the CEA (atomic energy commission) and the Institut Aérotechnique (aerotechnical institute) at St. Cyr.

The influence of the emission rate/wind speed ratio, the incidence of chimney height on the maximum concentration on the ground and the distance from the point of maximum concentration which is independent of the rate, have been studied by means of tests.

Experimentation has also contributed to the study of the height of chimneys, diffusion turbulence and the influence of buildings on diffusion.

3.2.5.3 Hydraulic model

Hydraulic models offer greater flexibility in adapting meteorological conditions. The hydraulic jet simulates to perfection the adiabatic conditions of atmosphere and wind (shape, velocity).

Colorimetric methods borrowed from colorimetric pH indicators used in chemistry give helpful visualization of the phenomena.

The scale of these models is 1/10.000.

Hydraulic models are quite appropriate for studying the upwards movement of a plume as a function of temperature, the outlet rate of effluents, wind velocity, temperature reversals, relief and the effect of a building.

3.2.5.4 Comments

- The validation of a dispersion model on mean distances requires knowledge of local weather forecasts based on a greater number of meteorological data.
- It is important to underline the fact that a model can lack accuracy for certain conditions (insufficient recordings, physical and chemical phenomena).
- At present, flow and dispersion models on the urban scale can be used for projects such as the location of an industrial zone and regulations concerning pollution.

However, there is a need for more research in this field especially as regards the forecasting of peak pollution.

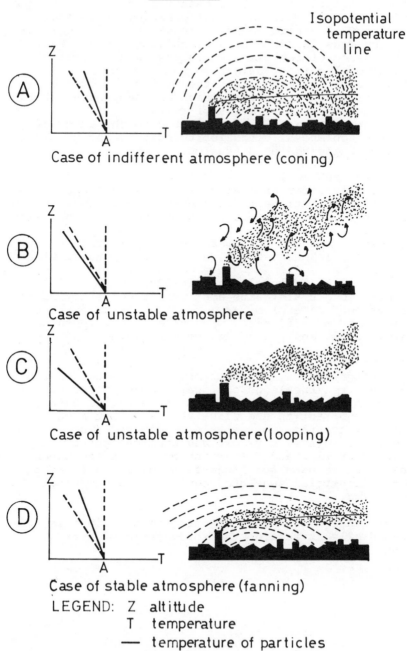

Case of indifferent atmosphere (coning)

Case of unstable atmosphere

Case of unstable atmosphere(looping)

Case of stable atmosphere (fanning)

LEGEND: Z   altitude
         T   temperature
         —— temperature of particles
         --- dry adiabatic

fig. 3.2.2.2. 1

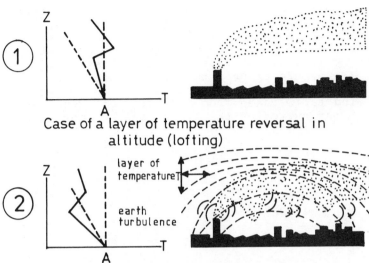

Case of a layer of temperature reversal in
altitude (lofting)

Case of a layer of temperature reversal in
altitude (smoke diffusion)

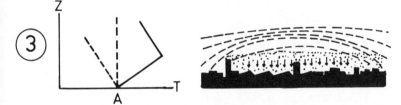

Case of strong dynamic stability with
temperature reversal beginning
at ground level and rising
above emission level

LEGEND: Z altitude
T temperature
— temperature of particles
--- dry adiabatic

fig. 3.2.2.3. 1

3.3 ATTACHMENT OF PARTICLES TO THE FACADE

3.3.1 Classification of particles

The ability of particles of dust to adhere to a substrate will differ according to their size, shape or nature.

In broad terms, the finest particles form unstable suspensions and are subjected to phenomena of coagulation, diffusion and turbulence, whereas particles over approximately 10 μm Ø are more exposed to gravity forces and tend to become deposited as sediment.

The following table taken from C.Carrie (cf. references) gives a brief survey of the types of dust found in the atmosphere.

Diameter of particles	Types of particles
0.0001 0.001	gaseous molecules
0.01	
	smoke (tobacco, coal, fuel oil, metallurgical industry
0.1	
	smoke, bacteria
1	
	bacteria fog mineral dust fly ash (coal fuel oil)
10	
	dust from mineral, cement and iron and steel industries fly ash fungi spores pollen mist
100	
	rain silt, rock debris dust fly ash
1000	

3.3.2 Cohesive forces between dust particles and adhesive forces to a substrate

3.3.2.1 Gravity forces
The largest particles are subjected to gravity forces F = mg. The friction forces "f" of the air prevent these particles from falling (fig. 3.3.2.1. 1).

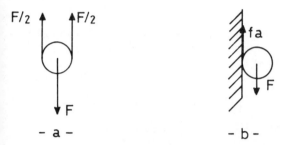

- a -   - b -

Others whose downward trajectory is more or less curved, come into contact with a surface, to which they adhere if the adhesive force exceeds the gravity force.
The velocity of the largest falling particles (500 to 1000 μm) can be defined by Newton's law :

$$v = \sqrt{\frac{8\,g\,d\,(m_1 - m_2)}{3\,m_2}}$$

g = gravity acceleration
m_1, m_2 = densities (dust,fluid)

For particles smaller than 50 μm, Stokes's law is considered.

$$v = \frac{d^2\,g\,(m_1 - m_2)}{18\eta}$$

η = dynamic viscosity of gas

3.3.2.2 Molecular forces
These forces are particularly predominant for the smallest particles, whose diameters are of the order of magnitude of the mean free path of molecules, that is to say about 0.01 μm.
C.Carrie and D.Morel (cf. references) describe the types of forces which affect the particles. These include polar, induction and dispersion forces (London forces).

93

The forces exerted between two close particles or between a particle and a surface are called van der Waals forces.

$$F = \frac{A}{12\,x^2} \cdot \frac{d_1 \cdot d_2}{d_1 + d_2}$$

d_1, d_2 = diameters of particles
x = distance between particles
A = constant

3.3.2.3 Electric forces

The particles in suspension in the air may be electrically charged. They are then attracted in varying degrees to a surface (glass, metal, plastic) or to each other or even repulsed if the surface or other particles are themselves charged (same sign).

The humidity of the air or the surface will decrease and even cancel these forces.

Forces of electrical origin also intervene in the phenomena of adhesion of particles to a surface.

3.3.2.4 Capillary forces

The incidence of water in the adhesion of dust on a surface is very important because of the development of surface tension forces at the solid/liquid/air interfaces. It is estimated that for the particles closest to the surface, these forces intervene when the relative humidity of the air exceeds 65%.

As an indication, the annual mean relative humidity in Paris is 76%, which shows the decisive nature of these forces in these conditions.

3.3.3 Effects of meteorological conditions : water - wind on the displacement of pollutants near a deposition surface

3.3.3.1 Atmospheric water

The important part played by the humidity of the air in the phenomenon of adhesion of particles to a surface has just been emphasized.

The presence of water as fine drops (mist) in the atmosphere, otherwise stable, facilitates coagulation, and as a direct consequence, sedimentation. This can result in the dust adhering to a building façade at a lower level.

The effect of rain, already referred to, is to capture the particles and precipitate them.

However, the joint action of rain and wind has the effect either of projecting the particles on to the surface or, by contrast, repelling them.

94

3.3.3.2 Wind

The contact of wind with a surface sets up turbulence.
This is illustrated by explanatory diagrams taken from
J.M. Huberty (cf. references) to differentiate the effects
of a light wind and a strong wind in contact with surface
relief.

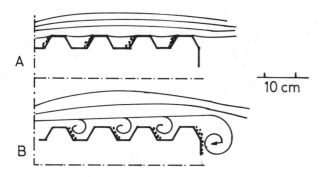

fig. 3.3.3.2. 1

In the case of a light wind (fig. 3.3.3.2. 1A) the
particles are mostly deposited on the windward face. On
the other hand, for a strong wind the eddies concentrate
the dust on the leeward surface (fig.3.3.3.2. 1B).

3.3.4 Dust condensation surfaces

In the book by R.T. Gratwick "Humidity in buildings -
causes and remedies", the phenomena of condensation of
dust on cold surfaces are described remarkably well.
   This description is appropriate here to contribute a
partial explanation of the displacement of dust particles
near a surface, assuming the influence of other parameters
such as wind, atmospheric water, and state of the atmosphere, to be neglible, however. (fig. 3.3.4. 1)
   These conditions can however probably be found locally
on a building.
   Gaseous molecules in the air are in a state of
continuous agitation ; consequently, any particle contained in the air is more or less exposed to this impact.
   We also know that there is greater molecular agitation
in hot air than in cold air, which gives it greater
lifting force.
   If the surface is colder than the air, a layer of
cooled air of varying thickness will be sandwiched between
the surface and the hot air.
   More intensive activity in the hot air which, moreover
constitutes a greater reserve of dust, has the effect of
directing towards the layer of cooled air more dust than
this layer discharges towards the warm air.

95

Progressively this layer of cooled air becomes
saturated with dust which settles on the cold surface.

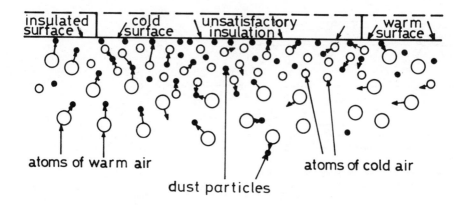

fig. 3.3.4. 1

In contrast, a surface warmer than the air heats the
contacting air layer, thereby forming a protective layer
against dust settling on the surface.
This is appreciable for dry air. Moreover, it is probable that the type of surface and more particularly the
colour and thereby the temperature of the surface, have a
direct influence on the development of these phenomena.
The presence of atmospheric water tends to encourage
the sedimentation of dust and consequently to minimize
dust condensation.
Nevertheless, these dust particles on which the water
molecules condense, even at a temperature higher than dew
point, can be driven on to the surface. They adhere there
all the better the higher the humidity content of the
surface.

References

(1) Carrie C., Morel D. and Fourquin J. 1975 Salissures
    des façades. Paris Eyrolles
(2) Gratwick R.T. Humidity in buildings - causes and
    remedies.
(3) Brebiou G. and Jullien M. 1973 Nuisances dues aux
    activités urbaines. Pollution et protection de l'air
    par L.Facy. Paris Guy Le Prat.
(4) Detrie J. 1969 La pollution atmosphérique. Paris

Dunod.
(5) Bourelly T. 1972 Interprétation des variations de la
    pollution atmosphérique en fonction des conditions
    météorologiques - Traitement des résultats.
    Etude C.S.T.B.
(6) Bourelly T. 1970 La pollution atmosphérique dans la
    région parisienne. Cahier 1038 C.S.T.B.
    aux quatre niveaux de la Tour Eiffel.
(7) Bourelly T. 1967 Etude de la pollution atmosphéri-
    que aux quatre niveau de la Tour Eiffel.

3.4 RAINFALL RUN-OFF

3.4.1. General description

Raindrops are guided towards buildings and building
components by three factors; the force of gravity, mass
forces and frictional forces acting against the air in the
flow zones which occur around the obstacles in the path of
the wind. In reality, the wind force, the wind direction
and the rain intensity vary markedly not only between
different showers of rain but also within the same shower
of rain. Consequently, it is not possible to carry out
any accurate calculations of the movements of the
raindrops.

The average vertical rate of fall for rain lies around
4-5 m/s and is, according to meteorological information,
roughly the same for most types of rainfall. The sizes of
the drops are also approximately the same from one shower
of rain to another.

The diameters usually lie between 2 and 5 mm. Using
these data as a basis, the rate of fall can be used to
determine roughly the forces with which the raindrops are,
on average, affected by air currents of different
direction and velocity.

In (1), (2) and (3) these calculations have been made.

3.4.1.1 Path followed by raindrops falling against buildings

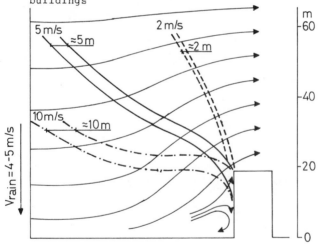

fig. 3.4.1.1. 1
⟶ wind trajectories
⎯⎯⎯ paths followed by raindrops which reach
‑ ‑ ‑ ‑ external wall surface in rain with
—··—·— different wind velocities

Figure 3.4.1.1 1 shows the path followed by raindrops falling towards a comparatively high, long building. The following can be seen from the figure.

(a) Less than half the quantity of rain which should pass through a free air cross-section of the same size as the building is caught by an external wall. This applies regardless of the wind force.
(b) The rain strikes mainly the top parts of the external wall. This has been confirmed by several measurements, including those published by Beijer. See fig.3.4.1.1 2. This figure also shows that there is no difference in the distribution between intensive and weak driving rain.

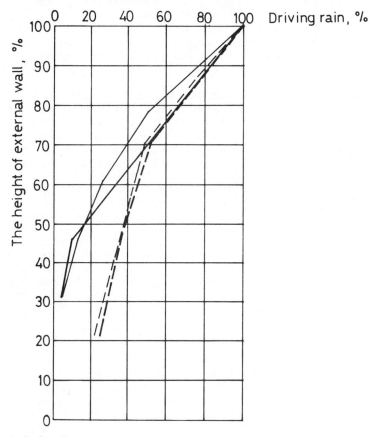

fig. 3.4.1.1. 2
Distribution of driving rain for intensive (coarse lines) and weak (fine lines) rain. The unbroken curves refer to a 5-storey building adjacent to a street in central Stockholm ; the broken lines refer to the gable wall of a comparatively free-standing 8-storey building.

(c) The raindrops move practically parallel to the lower sections of the external wall. This can also be seen from the soiling pattern that can be observed on many buildings. The width of the projecting part from the lower parts of an otherwise flat external wall need be no more than a centimetre or two to catch sufficient rain to give rise to clean-washed effects underneath.

In the case of external wall areas close to corners around which the air current is deflected horizontally, driving rain, naturally enough from what has been said in (b) and by analogy with the above, first strikes the outer areas. Raindrops are thrown somewhat further out from the air currents when deflected horizontally, when gravity does not directly affect the changes in direction, than is the case when they are deflected vertically.

In certain cases a dominating wind direction can play a decisive part regarding which sides of a building are reached by driving rain. The purely local conditions for the air currents are often of dominating significance here.

The size of the obstacles which the driving rain meets is of considerable importance for the manner in which the raindrops reach the obstacles. The shape of the air currents remains the same when the scale is changed from high-rise buildings down to obstacles not more than a decimetre in size. This means that the changes in direction of the air currents become more and more abrupt when the obstacle is reduced in size and that the time available for deflecting the raindrops from the path which the mass forces strive to give them becomes shorter and shorter.

Calculations results indicate that this scale effect should have a certain influence even for low buildings, particularly one-storey and, in certain cases, two-storey buildings. In these cases the driving rain reaches the lower parts of the external walls to a greater extent.

Ugly dirt patterns do not usually occur on such low buildings.

3.4.1.2 Path followed by raindrops falling against
         building components
The scale effect mentioned in subsection 3.4.1.1 above is very clear in the case of building components, even when they are fairly large. The raindrops are deflected in the manner illustrated in fig. 3.4.1.2. 1 when driving rain meets an obstacle with a width of 1 m projecting from a vertical surface. It can be seen from the figure that almost the entire width of the obstacle is reached by the driving rain.

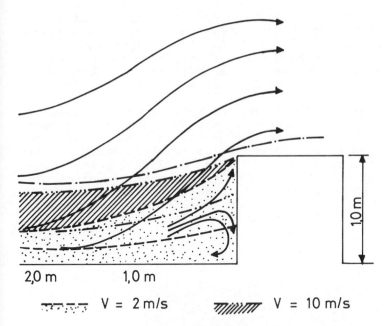

2,0 m   1,0 m

⟶⟶⟶⟶ V = 2 m/s   ⟫⟫⟫⟫ V = 10 m/s

fig. 3.4.1.2. 1.
Horizontal change of direction for raindrops when
approaching a 1 m wide obstacle which projects from a
vertical external wall.
The sectioned areas indicate for different wind velocities
the zone from which raindrops which occur in the air reach
the side surfaces of the obstacle.

   A study of the soiling patterns on external walls also
shows that the side surfaces of projecting components are
more cleanly washed by striking rain than are other lower
parts of the external wall and than are the front of the
components.
   The air currents are disturbed on the sheltered side of
an obstacle for a distance of about 5-7 times the height
of the obstacle. It should therefore be possible to
assume that, for example, pilasters at slightly more than
this distance from each other would receive the same
washing effects.

It can sometimes be observed that an external wall is kept clean immediately above flashing on top of mouldings or the like. See fig.3.4.1.2. 2. A corresponding phenomenon occurs to a far lesser extent on mouldings without flashing. The main reason for this is that the raindrops splash considerably less in the latter case.

The splashes which occur when individual drops strike different materials depend on the wetting properties of the material and also depend on whether (and increase markedly if) water already occurs on the surface in the form of a thin water film or drops which remain. Water of this type occurs fairly quickly on flashing during rain. Concrete surfaces absorb rainwater over a certain period of time. Consequently, the total quantity of water which splashes up on the wall becomes considerably less than in the case where flashing occurs.

The radius of action for splashes from sheet metal surfaces is surprisingly large. Observations of dirt behind several sheet metal surfaces with different clearances to the wall surface show that a certain washing effect can be obtained even when the clearance amounts to 15-20 cm.

References

(1) Beijer, O. (1980) Weathering on external walls of con-
    crete, Stockholm CBI-report 11-80, 70 pp.
(2) Sandberg, P.I. (1974) LTH Lund : Driving rain distri-
    bution over an infinitely long, high building : Compu-
    terized calculations (RILEM international Symposium,
    Rotterdam del. 1.1.2.)
(3) Cooper, R. (1974) Factors affecting the production of
    surface run-off from wind-driven rain. (Rilem inter-
    national symposium,, Rotterdam del. 1.1.2.)
(4) Beijer, O. and Johansson, A. (1976) Driving rain
    against external walls of concrete, Stockholm, CBI
    Research 7:76 92 pp.
(5) Beijer, O. (1977) Concrete walls and weathering. Rilem
    ASTM/CIB symposium on evaluation of the performance of
    external vertical surfaces of buildings, 29 aug-2 sept,
    Otaniemi, Finland, 10 pp.
Beijer, O. and Johansson, A. (1976) Water absorption in
    external wall surfaces of concrete. Stockholm, CBI-
    research 6:76, 49 pp.
Bielek, M. (1975) Joints and water movement on walls.
    Norges byggforskningsinstitut. Report 60, 39 pp.
Carrié, Morel, D. (1975) Salissures de facades. Paris
    (Eyrolles), 115 pp.
Clayton, J. (1975) Weathering - cleaning and restoration
    of concrete structures. Concrete 9, pp 43-44
Künsen, H. Schwarz, B. (1968) Driving Rain Measurements
    in Holzkirchen, Stuttgart (Institut für Technische
    Physik)
Robinson, G. Baker, M.C. (1975) Wind-driven rain and
    buildings. Ottawa 1975. National Research Council of
    Canada, Div. of Building Research. Technical Paper 445,
    50 pp.
Skárán, V. Kr. (1974) Some problems in industrial aero-
    dynamics : A literatur survey, Stockholm, Teknisska
    Högskolan, Inst för flygteknik. Aero rapport, BA 8,
    TRITA-FPT-007, 65 pp.
Huberty, J.M. (1980) Durabilité d'aspect des bétons
    apparents. Le viellissement de façades. Centre Scien-
    tifique et Technique de la construction. Brussel.

3.4.2 Influence of the absorption capacity of the external wall material

The raindrops which reach, for example, a wall surface are absorbed to different extents depending on the capillary suction capacity and the moisture content of the wall material. Brick walls and most plastered walls almost always completely absorb the quantities of water supplied by driving rain. The opposite applies to metal and glazed walls, for example. Rainfall run-off occurs very quickly after driving rain has started.

Beijer (4) has drawn up a calculation model for determining, with the aid of the capillary suction capacity of the wall material, how soon rainfall run-off is formed and how far down an external wall with a flat surface it reaches.

Fig 3.4.2. 1 shows an example of the magnitude of the rainfall run-off streams during a shower of rain, with a high and uniform driving rain intensity. It can be seen that it took about a quarter of an hour in this case before rainfall run-off began to appear and that the run-off then gradually increased as the wall material was moistened. Despite the fact that powerful driving rain continued for more than 1 hour, the run-off never reached ground level.

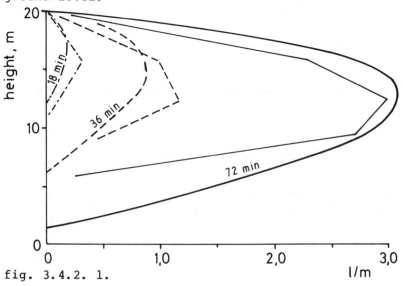

fig. 3.4.2. 1.

The manner in which the absorption capacity of the wall material according to the same calculation model affects the occurrence of rainfall run-off and the extent to which the run-off reaches ground level have been compiled in Fig.3.4.2. 2. This figure has been taken from (5).

104

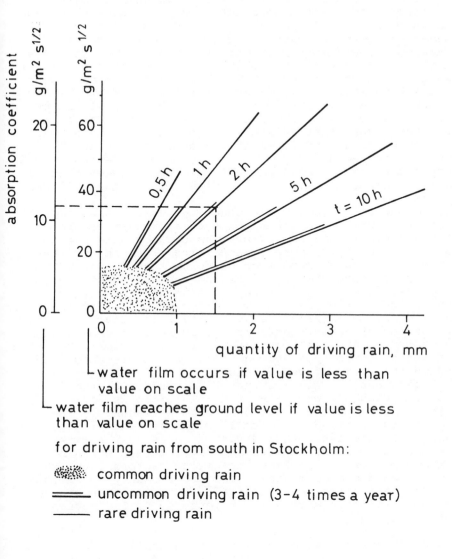

water film occurs if value is less than value on scale

water film reaches ground level if value is less than value on scale

for driving rain from south in Stockholm:

▨ common driving rain

══ uncommon driving rain (3–4 times a year)

── rare driving rain

fig. 3.4.2. 2
Influence of the absorption coefficient of an external
wall material on the rain run-off streams which result
from driving rain of different types and frequencies.
t indicates the duration of the driving rain in hours.
The figure shows that rain run-off rather often occurs on
concrete walls but usually affects only the upper parts of
the wall.

Different types of driving rain have been inserted in the figure and the frequency with which rain of this type occurs on average in the southern and central parts of Sweden has been indicated. It can be seen from the example in the figure that driving rain which contains a total of 1.5 mm water and which continues for a period of two hours (unusually powerful driving rain) gives rise to rainfall run-off on the external wall surface if the absorption coefficient of the wall material is less than about 35 g/m^2 $S^{\frac{1}{2}}$. At the same time, it can also be seen from the figure that rainfall run-off of this type does not reach the ground level if the same coefficient is greater than about 10 $g/m^2 S^{\frac{1}{2}}$. These observations can be compared with the fact that the absorption coefficient of facade concrete usually has a value lying between 10 and 40 g/m^2 $S^{\frac{1}{2}}$.

The outermost layer on a concrete surface cast against shuttering or on the top surface of a precast concrete unit contains a smaller proportion of aggregate per volumetric unit for purely geometrical reasons and thus contains a larger proportion of pores than do ground surfaces of concrete with dense aggregate. Consequently an outermost layer of this type has a larger capacity for capillary suction and for water retention. Grinding to a depth of no more than a few milimetres is enough for the proportion of cement paste in the surface to drop about one-third of the total surface. Grinding deeper than this and a high content of aggregate in the concrete further reduce the share of cement paste.

Consequently, rainfall run-off occurs on ground surfaces when the quantity of driving rain is lower. Rainfall run-off therefore occurs more frequently and reaches further. Since a smaller quantity of dirt is also encapsulated against the surface by gypsum, the water also remains cleaner and thus washes the concrete parts of the surface better. A combination of the factors involved, absorption and gypsum encapsulation, probably constitutes an explanation for the practical experience that ground concrete surfaces have a lower tendency to be unevenly soiled than unprocessed surfaces.

3.4.2.1 Paths and water quality of rainfall run-off
In (4) is also shown that rainfall run-off on concrete surfaces consists of a very thin layer - no more than a few tenths of a millimetre - and has a low velocity, of the order of up to 1 m/minute.

The detailed design of an external wall has a marked influence on the manner in which the rainfall run-off distributes itself over the surface. The soiling patterns clearly show this influence. The following general rules can be derived from such observations :

(a) The rainfall run-off is vertical. Lateral winds

have usually no more than an insignificant
influence
(b) The almost negligible depth of the rainfall run-off
means that even very small components affect its
distribution. On flat surfaces the rainfall
run-off often tends to break up into separate
streams. The location of these streams on the wall
surface is comparatively fixed. This breakdown
into separate streams takes place mainly on smooth
surfaces but can also occur on surfaces with, for
example, exposed aggregate.
(c) The rainfall run-off which driving rain causes
mainly on vertically projecting components often
forms concentrated streams with comparatively clean
water on the wall surface beneath the components.
Only slightly protruding window frames can give
rise to streams of this type. An example of this
effect consists of the soiling pattern reminiscent
of mustaches which occurs under insufficiently projecting window flashing which mainly catches water
from the side surfaces of window recesses.
Other components can also give rise to corresponding concentrations of rainfall run-off. See, for
example, the effect of the upper steel part on
the concrete screen in fig. 3.4.2.1. 1. The water
which reaches the steel parts follows to a certain
extent the vertical posts down towards the concrete
screen and gives rise to washed zones under the
fixture points.
(d) Narrow, projecting horizontal or moderately sloping
surfaces on low parts of external walls often give
rise to a fairly short cleanly washed vertical part
of the wall followed by a part which is soiled
slightly more than otherwise. See, for example,
fig. 3.4.2.1 2. The reason for this is probably
that the horizontal surface is soiled more rapidly
and also receives a far larger quantity of rainfall
per surface unit than adjoining vertical areas. The
water in the run-off which is formed on the
horizontal surface does not have time to get very
far before it is absorbed by the underlying
concrete surfaces since these unwashed surfaces are
comparatively dry. When the severely soiled water
is absorbed on a comparatively short section, it
gives rise to extra severe soiling directly
adjacent to cleanly washed areas above. Signs also
frequently cause this type of pattern. See, for
example, fig. 3.4.2.1. 3.
(e) Horizontal obstacles have a tendency to spread out
the water run-off which comes from vertical surfaces above. If the stream is to run on slightly or
moderately sloping surfaces, it needs a greater
depth than on the vertical surface above. It also

107

tends to spread out laterally.

In the same manner, rain run-off on the sides of columns which is, as has been indicated above, considerably more powerful than the rainfall run-off on the wall surfaces, tends to spread out on the top sides of the beams at points where projecting columns and beams cross, and bring with it the enriched dirt to the outsides of the beams adjacent to the column.

(f) The extremely thin film of rainfall run-off also runs along undersides.

The detailed design of drips is important. In the example in fig.3.4.2.1. 4 the drip under the horizontal window flashing had been completely omitted. If the water can be expected to follow the drip sideways, it is also usually necessary to take care of it at the ends of the drip.

(g) Vertical joints easily give rise to run-off streams which, unless they are dealt with, cause marked washing effects under the bottom termination of the joints.

By way of summary it can be said that the extremely thin run-off streams run in a fairly logical pattern mainly affected by gravity and the absorption capacity of the external wall material but more or less unaffected by lateral winds. The dirt content of the water increases further down. It also increases when the run-off meets horizontal obstacles on which dirt has accumulated during inter-rain periods.

Accumulation of dirt occurs to a greater extent on horizontal than vertical surfaces.

fig. 3.4.2.1. 1. The manner in which steel components are mounted can give rise to concentrated rain run-off.

fig. 3.4.2.1. 2. Soiling pattern caused by narrow horizontal surfaces under blind windows. A far larger quantity of driving rain per area unit strikes these surfaces compared with the vertical surfaces. Rain run-off is formed as a result and carries the accumulated dirt with it on the horizontal surface. The absolute quantity of water is, however, small.

fig. 3.4.2.1. 3 Horizontal surfaces projecting from an
external wall give rise to moderate streams of very dirty
water.
This picture is from Carrie and Morel (1975). It shows a
plastered external wall. Similar soiling pattern occurs
on concrete walls.

Chapter Four

Biological soiling

4.1 INTRODUCTION

The deterioration of artistic stone masterpieces has been
studied for many years with the aim of improving the processes used for preserving this historical heritage. However, quite often the investigation undertaken only concerns very specific cases - there is little evidence of a
basic study as regards the different types of soiling
found on building façades.

Apart from rare exceptions, research scientists have
especially left aside soiling by plant microorganisms
which develop on different types of façade materials.

Knowledge of bacterial action in deteriorating concrete
sewer networks and the damage caused to concrete
structures such as cooling towers (21, 37) of dams (16) is
well established, as is that of other microorganisms in
modifying the appearance of concrete or masonry walls
(31).

There are also numerous research reports on the
deterioration of stone monuments (4, 5, 14, 15, 20, 28,
41, 43). Although all these results cannot be directly
applicable to the other materials used in façades; some
of them give interesting indications, however.

In the case of concrete- or cement-rendered façades, a
recent study (2, 29) has shown that the microscopic flora
in soiling investigated in a city subjected to light pollution, located beside the Atlantic Ocean, consisted of
three different types of organism :

a. bacteria, most often sulphooxidizing and heterotro-
    phic, that is to say oxidizing sulphur and growing
    on a purely mineral or possibly organic substrate;
b. microscopic algae, either of the blue or green
    type. They are generally autotrophic and can
    develop on purely mineral substrates and even in
    certain cases in areas with little light;
c. fungi, heterotrophic organisms generally developing
    on substrates containing organic matter.

In fact, these three types of microorganism most often
live in colonies, varying in composition according to
conditions. We have evidence that soiling can have a
general aspect of black, red or green colour (2).

This type of coloured soiling is not restricted to the

area in which the study was undertaken, since the same type of soiling has been found in other areas of the same country (29) or in other countries at some distance (25, 35, 42).

When soiling is well established over time, a modification in the flora is found. In fact, lichens - another type of plant life-can form from certain algae and fungi. Their exact mode of formation is still unknown today, but we know they are made up of algae and fungi existing in symbiosis.

At more advanced stages of deterioration, other types of plant life may be involved, such as moss, ferns and even higher botanical species.

Such plant life corresponds to quite complex evolution stages, so it is preferable to limit our scope to the first organisms referred to, that is to say bacteria, algae and fungi, without however forgetting lichens which present certain interesting characteristics.

This proliferation of microorganisms is not confined only to concrete façades or cement-based renderings - there is evidence of bacterial deterioration to asbestoscement units (21, 27). It is also certain that plastics are sensitive to certain bacteria and fungi (24, 26) even if this type of deterioration has not always been made obvious (33, 34). It is also well known that natural slates are often covered by green algae, lichens and moss (13). Synthetic asbestos-cement slates increasingly undergo the same defacement, as do most roofing materials, including baked clay or concrete tiles. Other studies also show that in certain conditions , asphalt road coatings can deteriorate owing to the presence of fungi (22, 23). Finally, paints and thick plastic coatings are also sensitive to the proliferation of fungi and other organisms (29, 32).

4.2 VIABLE PARTICLES IN THE AIR

Up to high altitudes (28 000 m), the atmosphere naturally contains viable particles, that is to say microorganisms such as bacteria, algae and fungi in the shape of fragments or spores. It also contains large numbers of pollen grains, animal life, microscopic creatures such as protozoa or larger forms of insect life.

4.2.1 Bacteria

According to the conditions, bacteria which are unicellular organisms exist as living cells, isolated or grouped in colonies, or as thermoresistant spores.

These are microorganisms as found everywhere, but it seems that the bacteria transported by the atmosphere originally come from soil or water, or more specifically from filtering stations.

These organisms generally vary in size from 0.3 to 15 micron and are found in variable quantities according to sampling locations. For example, the air sampled at the summit of Mont Blanc contains only 4 to 11 bacteria per cubic metre, where as measurements taken in New York show that the air contains from 100 to 1000 bacteria per cubic metre. The quantity of bacteria is much lower at sea and air sampled over an ocean will only contain 4 - 5 bacteria for 10 cubic metres. The number of bacteria per cubic metre of air also varies with latitude, the concentration being much lower in polar air than in tropical air.

It has also been shown that bacteria can be transported in the air in small drops of liquid and that the concentration of microorganisms in these drops can be much higher than that in the liquid from which they came (from the sea).

4.2.2 Algae

These are present in the air as cysts or spores, that is, a vegetative life form.

Their number per cubic metre is very variable from one point of sampling to another (from 1 to 200 colonies per cubic metre of air) and their size is generally about 0.5 micron.

They mainly originate in the soil and become air-borne with dust.

Their resistance to desiccation is high and colonies can develop in the presence of sufficient humidity after more than a year passed resisting in dry conditions.

Algae present in the air are of very numerous kinds. The composition of the algae flora in the air varies with the algae population of soils from which it originates and also with meteorological conditions.

4.2.3 Fungi

Fungi are very frequently present in soil and dust. Many species breed from spores adapted to air-borne dispersion. It has also been shown that cereal diseases due to fungi (blight, mildew) are propagated over very long distances by wind-borne spores.

The concentration of fungi spores in the air decreases with increasing altitude, but for one particular area, they also vary with the seasons. Research has revealed that if fungi spores are permanently present in the air at an altitude of 45 m, their concentration is at a maximum in July and August (from 1800 to 25000 spores per cubic metre) and a minimum during the winter months (from 170 to

700 spores per cubic metre). However, in most cases, these spores are not all viable and only 30% - 60% of them will turn into colonies, even if viability rates up to 90% have been observed by some research scientists.

The sizes of fungi spores vary with the species from 3 to 100 microns, which corresponds to the size of pollen grains. It is therefore quite obvious that only the species with the smallest spores can be transported over long distances.

4.2.4 Specific comments

The different microorganisms may also be found in the atmosphere grouped in larger particles or attached to other elements such as dust or pollen grains. This phenomenon helps to explain the simultaneous propagation of several microorganisms.

As regards lichens - microorganisms formed by the symbiosis of a fungus and an alga - research undertaken up to the present shows that they can breed either in a vegetative way by fragmentation of the thallus, which corresponds to linking of fungus pieces or hyphae with algae cells, or from particular organs (isidia and soredia) which bring about the dispersion of fungus or algae cell hyphae by outgrowth. Moreover, the lichen fungus can produce its own spores and contribute to the spreading of fungus spores. In this last case, new lichens can be formed by the capture of algae cells.

Since lichen reproduction or breeding elements are not easy to differentiate from those of fungi or algae, this offers an explanation of the absence of this type of element in almost all the results of analysis of air-borne particles, even if there is reason to believe that they are effectively present.

Research has shown that the types of microorganism present in the air are a function of its quality, i.e. the microorganisms vary according to the degree of pollution (3). It is therefore obvious that the population of microorganisms found on façades will be different according to the areas in which the buildings concerned are located (rural areas, suburbs of light pollution, urban zones with or without heavy pollution).

Finally, as regards the attachment of air-borne microorganisms to façades, although no precise information is available from research on this subject, there is reason to believe that it is either linked to electrostatic phenomena, to the presence of very hygroscopic mossy sheaths which surround certain living particles and which are extremely adnate (14) or perhaps to surface tension forces developed by the presence of varying quantities of free water at the surface of the material or in the pores opening on to the surface. In practice, for most building

materials used, their porosity and therefore their water
retention capacity (1) is sufficient to support the existence and growth of the microorganisms encountered in
soiling.

4.3 CONDITIONS FOR MICROORGANISM DEVELOPMENT

In a general way, the development of microorganisms
depends on quite a large number of factors which are
interdependent. It is very difficult to define precise
and simple development criteria.
   However, the four main factors are clearly the pH,
water, temperature and light.

4.3.1 pH

pH action on most microorganisms depends on their toleran-
ce of this factor. These organisms can be classified into
acidiphiles, neutrophiles and basiphiles (12), but there
also exist indifferent microorganisms which develop normally in a wide range of pH, such as bacteria whose growth
is satisfactory between pH values of 6.0 and 9.0 and numerous fungi which tolerate pH values from 2.0 to 11.0 (30).
   Most algae found on concrete façades or cement-based
rederings can normally develop at a pH of 8.0 (17).

4.3.2 Water

Water is undoubtedly the element essential to the life of
microorganisms and therefore to their growth, but it also
intervenes indirectly through the part it plays in gaseous
exchanges and the transfer of nutritive substances.
   Although we have no precise information on the hydric
development conditions of microorganisms on the surface of
materials, by analogy with the results known in the field
of soil microbiology, we can assume that if water tension
reaches 40 MPa, all microbial activity ceases (12). For
lower tensions, activity varies according to the microorganisms and seems to reach a maximum at a tension around
0.001 MPa for most organisms involved in façade soiling.
Beyond this point it decreases again.
   This corresponds to very high relative humidities close
to saturation (at least 95% RH). It must be noted however
that, taking into account the temperature differences
existing between a wall and the environment, the relative
humidity of the boundary layer near a surface can be quite
different from that of the surrounding air. This helps to
explain the development of microorganisms in apparently
unfavourable conditions (39).
   Moreover, if desiccation is such that the limit thres-

hold of microorganism activity is exceeded, there may be extensive destruction of the microflora, but there will not be complete sterilization of the medium because, even in the case of very fragile microorganisms, certain cells or microcolonies can survive a long time in various forms.

4.3.3 Temperature

Generally speaking, for each species of microorganism there is an optimal temperature for growth and a range between a minimum and a maximum outside of which development is impossible. The optimal temperature generally lies between 25 and 40° for most bacteria. Fungi are seen to be very tolerant as regards temperature, whereas algae are very sensitive to it even if their requirements vary considerably from one species to another.

Resistance to high temperatures (above 50°C) or low temperatures (below 0°C) appreciably varies from one type of microorganism to another. Some can resist temperatures above 90°C, particularly all the species which can form spores or cysts. Slow freezing usually causes destruction of microorganisms, but some can survive a long time at quite low temperatures.

Most microorganisms also show a remarkable ability to adapt to thermal and hydric cycles. These cycles are quite obviously linked, and are sometimes necessary for development in cases where they encourage sporulation.

4.3.4 Light

Photosynthetic organisms such as algae and certain bacteria need light and carbon dioxide to be able to develop normally.

Non-photosynthetic microorganisms, on the other hand, are generally accepted to undergo biocide action from solar radiation, although this has not been checked systematically.

4.3.5 Nutritive conditions

Growth is also linked to the presence in the medium of different elements necessary to the microorganisms metabolism and also to the intervention of development inhibitors in certain cases.

There is little precise information available on this subject as regards the development of bacteria, algae and fungi, but for lichens it has been shown that their presence in certain places is a relatively precise indication of atmospheric pollution (8, 9, 10, 40).

116

4.4 EFFECTS OF MICROORGANISMS ON SUBSTRATES

The microorganisms on façades can modify the substrate medium by their presence and mode of existence. Indeed, these living microorganisms draw the elements necessary for their nutrition from their surroundings and reject waste in variable quantities which can cause deterioration in the substrate material.

4.4.1 Effects of bacteria

Research has shown that certain types of bacteria - sulphooxidizing and sulphate-reducing bacteria - exert a corrosive action on stone (6, 19).

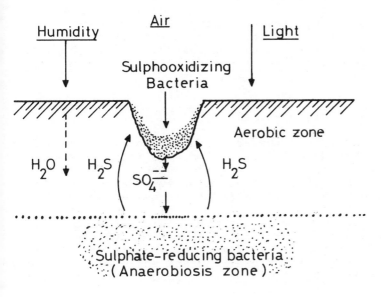

fig. 4.4.1. 1.

From the sulphur stored in their reserve, the autotrophic sulphooxidizing bacteria which develop on the surface of materials, excrete sulphuric acid. Reacting with the substrate, this acid causes the formation of soluble sulphates which diffuse inside porous materials. The sulphate-reducing bacteria, which live in anaerobiosis inside pores at a certain distance from the surface, will then use these sulphates and break them down, giving off hydrogen sulphide. This will in turn migrate towards the exterior and is assimilated by the sulphooxidizing bacteria to reconstitute their sulphur reserves. This forms the "sulphur biological cycle" - the well-known natural phenomenon which has been shown to be responsible for stone surfaces flaking and also for deterioration of concrete façades and cement-based mortar renderings.

In the same way, but in a more limited number of cases, research has revealed the nitrogen biological cycle mechanism which leads to the formation of nitric acid and nitrates drawn from the nitrogen in the air or from organic nitrogen from ammonium ions assimilated by different bacteria. This phenomenon has been recognized as being responsible for the alveola deterioration of stones (15) and degrading of asbestos-cement units in certain cases (21, 27).

4.4.2 Effects of algae

The algae found in soiling are mainly autotropic, which is to say that they do not rely on the substrate for food. However, they secrete organic acids which can dissolve the calcium carbonate of limestone, concrete and mortar and they can absorb this calcium for their own metabolism (38).

Apart from this chemical effect, algae can also act on their substrate by the insertion of cells into the pores. During alternating periods of wetting and drying, these cells will be the seat of swelling and shrinkage which can have a mechanical influence on pores and lead to cracking of the substrate.

Moreover, it is important to note that most microscopic algae met with are surrounded by a hygroscopic, mossy sheath which retains water, so the presence of an algae covering on the surface of a porous material can bring about a rise in the humidity content of the underlying material.

Finally, these mossy sheaths subsist after the destruction of algae by drying or by temperature variations beyond the tolerable limits. They then serve as a basic nutrition for bacteria or fungi.

4.4.3 Effects of fungi

Fungi are heterotropic organisms which mostly use organic matter for their nutrition. But they can also secrete organic acids which may attack the substrate. They are above all characterized by their production of very power-ful enzymes capable of degrading the carbonate chains so as to free carbon which they will oxidize and assimilate for growth.

It is this phenomenon which is responsible for the de-terioration of paints and thick plastic coatings by fungi. This leads to decomposition of paint films which lose their properties and finally disintegrate.

4.4.4 Effects of lichens

Certain lichens found on façades are simply attached to the surface of their supporting medium and therefore exert no mechanical action on the façade materials. But in very numerous cases the thallus penetrate several millimetres into the substrates, not only through the cracks and exis-ting pores, but also by attacking the material.

Lichens secrete carbon dioxide and acid products which can react with the carbonates by forming soluble salts such as for example acid calcium carbonate. Moreover, other substances produced by these organisms, lichenic substances, have in their molecules the -OH and -COOH groups which can, through the phenomenon of chelation, take over metallic atoms such as Ca and Mg. In this case, the minerals are destroyed without there being salt formation.

4.4.5 Propagation of soiling of biological origin (29)

Soiling caused by living microorganisms has a different effect on the surfaces affected than disorders due solely to mineral dust.

Although the bacteria cannot be seen with the naked eye, the soiling due to algae or fungi is quite visible and even striking.

Fungi are characterized by concentric growth from an original central point, whereas soiling comprising mainly algae accompanied by bacteria and fungi, propagates in the shape of a conical trail falling from a high point often located near an edge (the top of a wall for example). In this second case, run-off water constitutes the real propagation vector.

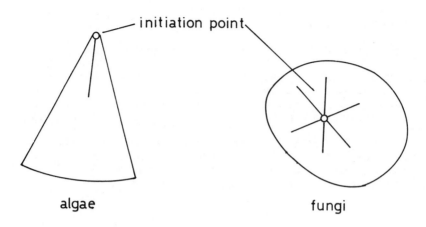

initiation point

algae   fungi

fig. 4.4.5. 1

In the case of obstacles projecting out from vertical
surfaces, it is also easy to distinguish soiling of biological origin : by washing the surface, the rain tends to
clean better either side of the obstacle. Mineral soiling
will therefore tend to persist at points with less washing
forming in this way a kind of "beard" under the obstacle,
whereas soiling consisting of microorganisms will form
"moustaches" each side of this same obstacle.

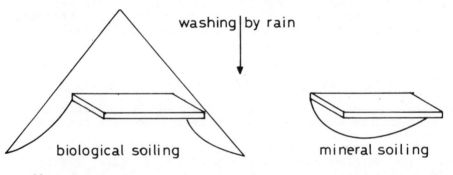

washing by rain

biological soiling   mineral soiling

fig. 4.4.5. 2

Finally, we have already mentioned the importance of
the displacement of microorganisms by wind and atmosphere.
It is for this reason that we can observe geographical
areas where certain types of soiling persist, easily noticeable through their specific colours (29, 35), areas
affected by relief and the direction of prevailing winds.

References

(1)   Ashton, H.E. and Sereda, P.J. (1982) Environment,
      microenvironment and durability of building materi-
      als. Durability of building materials; 1, 1, 49-65
(2)   A.U.C.U.B.E. (Agence d'Urbanisme de la Communauté
      Urbaine de Brest et de son Environnement),I.N.S.A. de
      Rennes (1983). Une prolifération mal contenue : les
      microorganismes sur les façades des constructions.
      Rapport AUCUBE; 53, 15 p.
(3)   Babich, H. and Stotzky, G. (1978) Atmospheric pollu-
      tion and microorganisms. Am.Soc.Microbiol.News, 44,
      10, 547-550.
(4)   Barcellona Vero, L. and Montesila, M. (1976) Mise en
      évidence de l'activité des thiobacilles dans les
      altérations des pierres à Rome.   Identification de
      certaines souches Colloque International UNESCO-
      RILEM, Paris 4.1, 13 p.
(5)   Bassi, M. and Chiatante, D. (1976) The role of pigeon
      excrement in stone biodeterioration.   International
      Biodeterioration Bulletin, 12, 3, 73-79.
 (6) Chantereau, J. (1980) Corrosion bactérienne. Bacté-
      ries de la corrosion. Technique et documentation,
      262 p.
(7)   Censoni, A.L.Z.,   Bettini, C., Giacobini, C. and
      Mandrioli,P. (1979) Aerobiological research in
      enclosed space of historical and artistic interest.
      Umwelt Bundes Amt Berichte, 5, 434-438.
(8)   Deruelle, S. (1983) Effects à long terme de
      la pollution atmosphérique sur les lichens, Sixième
      Congrès Mondial pour la Qualité de l'air, UIAPPA,
      3, 533-540.
(9)   Deruelle, S. and Lallemant R. (1983) Les lichens
      témoins de la pollution. Thèmes Vuibert Université
      Biologie, Vuibert, 108 p.
(10) Deruelle, S., Lallemant, R. and Roux, C. (1979) La
      végétation lichénique de la Basilique N.D. de
      l'Epine (Marne), Documents Phytosociologiques, 4, 217
      - 234 and 234a - 234e.
(11) Detrie, J.P. and Jarrault, P. (1968) Pollution
      d'origine végétale; Extrait de : Les industries,
      Leurs productions, Leurs Nuisances. La Pollution
      Atmosphérique, Dunod, 6-12.
(12) Dommergues, Y. and Mangenot, F. (1970) Ecologie
      microbienne du sol. Masson et Cie, 796 p.
(13) Drevet, J.P. (1980) La végétation parasite
      sur les couvertures en ardoise. Société Ardoisière
      de l'Anjou.
(14) Dupuy, P., Trotet, G.aand Grossin, F. (1975) Protec-
      tion des monuments contre les cyanophycées en milieu
      abrité et humide. The Conservation of Stone, Procee-
      dings of the International Symposium, Bologne,
      205-219

(15) Faugeres, J.G. (1978) Altération et traitement des pierres calcaires en oeuvre. Une étude concrète du Laboratoire Municipal de Bordeaux. Cahiers Techniques du Moniteur, 18, 9-17.

(16) Gore, P.S. and Unnithan, R.V. (1977) Thiobacilli from Cochin backwaters and their oxidative and corrosive activities. Indian J.Mar.Sci. 6, 2, 170-172.

(17) Grant, C. and Bravery, A.F. (1981) Laboratory evaluation of algicidal biocides for use on constructional materials. 1. An assessment of some current test methods, International Biodeterioration Bulletin, 17, 4, 113-123.

(18) Jacobson A.R. and Morris S.C. (1977) The primary air pollutants. Viable particulates, their occurrence, sources and effects. (Abstract) Stern A.C., Air Pollutants, their transformation and transport, Academic Press In., Troisieme Edition.

(19) Jaton, C. (1974) Attaque des pierres calcaires et des bétons, (Abstract) Association des Ingenieurs en Anticorrosion, Dégradaton microbienne des matériaux Ed. Technip., 41-51.

(20) Jaton, C. and Orial, G. (1978), Ecologie microbienne de murs expérimentaux en pierre calcaire et traitement indirect avec des hydrofuges. Colloque International UNESCO-RILEM 4,5, 30 p.

(21) Kaltwasser, H. (1976) Destruction of concrete by nitrification. European Journal of Applied Microbiology, 3, 3, 185-192.

(22) Khimerik, T.Y. (1979) Studies of fungi effect on the durability of asphalt coverings (in Ukrainian), Mikrobiol. Zh., 41, 3, 245-251.

(23) Khimerik, T.Y. and Koval, E.Z. (1977) Recherches sur la résistance aux champignons de quelques matériaux de construction de routes (in Ukrainian), Mikrobiol. Zh, 39, 1, 84-87.

(24) Kilbertus, G (1978) Premiers stades de la dégradation microbiologique des PVC. Etude électronique, Mater. Org., 13, 2, 89-100.

(25) Marathe, K.V. and Sontakke, S.D. (1977) Observation of some wall algae. Bat. Nagpur, 8, 1-4, 31-34.

(26) Naplekova, N.N. and Abramova, N.F. (1978). Dégradation microbiologique des plastiques (in Russian). Izv. Sib. Obk. Akad. Nauk, Ser. Biol. Nauk, 15, 42-47.

(27) Novoty, J., Wasserbauer, R. and Zadak, Z. (1972) Influence du facteur biologique sur la destruction des couvertures en amiante-ciment des étables. Premier Colloque International sur laDétérioration des Pierres en Oevre, La Rochelle, 155-156.

(28) Paleni, A. and Curri, S.B. (1972) L'agression des algues et des lichens sur les pierres et les moyens pour la combattre. Premier colloque International sur la Détérioration des Pierres en Oeuvrela Rochelle

157-166.

(29) Perrichet, A. (1984) Développement de microorganismes à la surface des bétons et enduits. Materiaux et Constructions, 17, 98, 173-177.

(30) Pochon, J. and De Barjac, H. (1958) Traité de microbiologie des sols, Dunod, 685 p.

(31) (1956) Quelques aspects de la corrosion biologique des mortiers et bétons. Revue des Matériaux de Construction, 485, 50-52.

(32) Rabate, H. (1975) La peinture dans la contruction. Environnement, pollution et souillures biologiques des revétements par peintures et préparations assimilées et par feuilles de matières plastiques. Revue Technique du Bâtiment et de la Construction

Industralisée, 22, 47, 89-97.

(33) Rechner, L. (1971) Contribution à l'étude du vieillissement naturel des matières plastiques utilisables dans la construction (First Part). Cahiers du CSTB, 1077, 45 p.

(34) Rechner, L. Vieillissement naturel des matières plastiques utilisables dans la construction. Cahiers du CSTB, 1453, 49 p.

(35) Ryan, N.M. (1983) Algal growth on cementitious surfaces. Pilotstudy ; National Institute for Physical Planning and Construction Research, Dublin, 9 p.

(36) Shafiee, A. and Rahmani, T. (1978) Atmospheric mold spores in Teheran. Ann. Allergy, 40, 2, 138-142

(37) Taylor, C.B. and Hutchinson G.G. (1947) Corrosion of concrete caused by sulfuroxidising bacteria.J. Soc. Chem. Industry, 66, 54-57

(38) Trotet, G., Dupuy P. and Crossin F. (1972) Sur une nuisance biologique provoquée par les cyanophycées ; premier colloque international sur la détérioration des pierres en oeuvre, La Rochelle, 167-170

(39) Urquhart, D.C.M. (1982) Condensation et développement des moisissures dans les maisons. Batiment International, 10, 2, 88-99

(40) van Haluwyn, C. and Lerond, M. (1983) Lichens : Végétaux tests de la pollution atmosphérique. Sixième Congrès Mondial pour la Qualité de l'air, UIAPPA, 3, 469-474

(41) Verona, O. and Gherarducci, V. (1977) Alcuni aspetti del deterioramento delle opere monumental della cita di Pisa. Riv. Ital. Ing., 37, 2-3, 83-91

(42) Wee, Y.C. and Lee, K.B. (1980), Proliferation of algae on surfaces of buildings in Singapore. International Biodeterioration Bulletin, 16, 4, 113-117

(43) Wood, P.A. and MacRae, I.C. (1972) Microbial activity in sandstone deterioration. Biodeterioration Bulletin, 8, 1, 25-27.

Chapter Five

Cleaning of soiled façades

5.1 INTRODUCTION

Cleaning, restoration and conservation of the façade are closely related to each other and are not to be considered separately.

Cleaning means in the strict sense of the word that dirt is being removed from the façade without changing the surface texture of the surface material itself.

The aim of restoring the façade is to bring the material properties, among which the outward appearance, back to its initial state, so to enable the façade to perform its functions effectively.

The purpose of conservation is to retard as much as possible the modification of the structure of the original or of the cleaned surface due to atmospheric influences.

With soiled façades the appearance is to be considered of primary importance. It appears in practice that in most cases cleaning also involves the modification of the surface structure. With the cleaning of stones by applying chemicals or by blasting, the surface structure of the material will be changed.

Moreover the surface has already been damaged during the soiling process, so that the initial structure no longer exists even before the removal of the dirt.

The removal of the outer skin can therefore be considered as a side-effect of the cleaning process.

Façade restoration for the purpose of returning the material properties to a known earlier state frequently cannot take place if the dirt has not been removed beforehand.

Advanced soiling is not simply an aesthetic problem, because dirt can also be a major cause of failure. The accumulation of dirt will stimulate the damaging crystallization of salts beneath the surface and the harmful forming of acid deposits. Very few porous stones can withstand the pressure exerted by crystals growing beneath the surface.

Before it is possible to proceeed to conservation, the façade has to be cleaned beforehand otherwise the conservation agents cannot enter the porous surface of the soiled material.

It is therefore advisable to consider carefully the main purposes and also the side-effects before arriving at any decision to clean.

The following purposes for cleaning can be distinguished :

(a) Cleaning for purely aestetic reasons.
(b) Cleaning with the main purpose to remove potent aggressive deposits from the surface of the façade, which will cause decay of the façade material. Second aim may be the view to improve the appearance.
(c) Cleaning as a preliminary treatment before the applying of conservation agents or paint.

In all cases the following factors must be considered in order to come to a decision on cleaning :

(a) The appearance of the façade before cleaning and the finished appearance to be expected after cleaning.
(b) The direct damage to buildings, people and environment caused by the cleaning procedure.
(c) The positive or negative effects on the durability of the façade.
(d) The economic aspects.

Concerning the cleaning methods a choice can be made between the following methods :

(a) Washing.
(b) Chemical cleaning.
(c) Mechanical cleaning.

This classification of the cleaning methods in three groups according to the nature of the detachment of the dirt deposits is somewhat arbitrary.
In practice commonly applied cleaning systems are frequently a combination of 2 or all 3 of these methods.
The selection of a method is governed by :

(a) The motivations as above-mentioned.
(b) The characteristics of the façade which are of influence on the soiling process namely : the properties of the façade material, material deficiences, architectural details, incorrect designs etc.
Under the same prevailing environment conditions different kinds of soiling will take place according to the nature of the façade material, which reacts differently upon the soiling actions.
(c) Prevailing atmosferic conditions, local microclimate, presence of pollution, locality and nature of the building.
(d) Character and causes of the soiling process.
(e) Nature, expected results and economic aspects of the cleaning methods to be considered.

5.2 SOME PHYSICAL PRINCIPLES OF CLEANING

Cleaning can be divided into two operations :

1. The detachment of the particles from the substrate.
2. The transport of the detached particles from the surface.

ad.1. The detachment of the dirt particles is based on different processes. Which process come into consideration depends on the type of bonding between particles and substrate.
To be distinguished are :
a) Action of water or steam.
Water with or without a surface active agent or steam may wet the various dry particles. The water layer surrounding the particles diminishes the mutual attraction between the particles and surface of the façade. The dirt now becomes more or less plastic. The action can be accelerated by adding a surface active agent. As some very fine dry clay like dust is nearly impermeable to water, it may take some time for the water to penetrate.
b) Dissolution of the substrate.
Water can sometimes dissolve the outer skin of the surface of the façade. In pure rainwater hardened cement paste can slightly dissolve. The bonding of the particle to the substrate is then destroyed. The use of acids is based on the same principle.
c) Dissolution of the cementing products of the dirt. As particles are often fixed to the surface by plaster ($CaSO_4 . 2H_2O$), that has been formed by the reaction between the SO_2 from the air and the lime of the substrate it is possible to obtain a cleaning effect from the dissolution of the plaster by water. Plaster is partly soluble in water : 2 g/l.
d) Mechanical process such as dry or wet blasting, spinning off, brushing etc.
e) Saponification of fats by alkalis or acids. This is also based on a chemical reaction.

ad.2. The transport medium can be air or water.
In some cases the detachment and the transport is effected by one and the same process.

5.3 CLEANING METHODS

5.3.1 Washing

Water plays a prominent role in the cleaning process due to its ability to soften, dissolve and physically dislodge deposits.

As a matter of fact two phases are to be recognized with water cleaning : first, the dirt will be softened by spraying or sprinkling (gently) onto the surface. Presoftening will facilitate washing away with water. Emulsifying agents may be applied and mixed with the water, if the dirt contains grease and oil compounds.

The next phase is the detaching of the softened dirt from the underlying surface, the loosening of the bond between them.

This process can be done manually or mechanically by brushing the dirt off or by directing a stream of water onto the surface.

The pressure of the water can be varied considerably from low to high. In the case of low pressure it will be done mostly in combination with handbrushing. The higher the water pressure the more effective the cleaning but the greater the damaging effect to the surface. The water can also be mixed with grit in order to increase the abrasive action.

The mean of transport for the removal of the loosened dirt is here again the water.

It is clear that the method, where the dirt is pre-softened by gentle spraying and then loosened by brushing, has only a slight abrasing effect on the underlying surface. This method is therefore very useful for cleaning up soft and only moderately dirtied façades. This occurs mostly with concrete façades.

Water jetting at high pressure or wet grit blasting is appropriate for the removal of heavy, tough soiling on hard and dense surfaces.

The main general disadvantage with wet cleaning is that it may damage the façade through unwanted soaking.

It will leave very wet parts behind, long after the cleaning is carried out, and this may cause harmful effects such as efflorescence, change in colour, organic growths etc.

The loosening of the pre-softened dirt is also to be accomplished by directing heated steam of low pressure towards the façade.

Steam cleaning does not damage the stone unless it is very soft. It is a useful method of loosening oily, greasy or tarry deposits, but it is not favoured because it is slow and not particularly effective in removing dirt. As a rule the façade should be cleaned by the least damaging method.

127

5.3.2 Chemical methods

If the use of only water is not sufficient enough to remove the dirt, chemicals will be added to the water to assist the cleaning. Chemical agents assist in the removal of dirt due to their ability to react chemically with the dirt, to dissolve it or to destroy the cohesion by saponification.

Chemicals are generally used to soften and loosen the dirt before washing or other general cleaning process.

The method of application is to wet thoroughly the surface, spray or brush on a minimum of the chemical agent, allow it to act for a short period and then rinse it off.

Acids :

The most common acids used for cleaning are :

Hydrofluoric acid (HF) and Ammonium bifluoride (NH_4HF_2) As an alternative to HF: Hydrochloric acid (HCl), Phosphoric acid (H_3PO_4), Sulphuric acid (H_2SO_4), Oxalic acid $(COOH)_2$, Acetic acid (CH_3COOH), Formic acid (HCOOH).

The last three of these acids are weaker acids. There does not exist one universal acid that can be used in all circumstances but for each substrate only some acids are fit.

Acids will transform the dirt from a fixed, non-transportable state to a movable transportable state and thus facilitate washing away with water. The removal of the dislodged dirt will be done by pressure-water spraying.

Attention must be given to the following terms if acid is applied in the cleaning operation :

(a) The façade must be saturated with water prior to the application of the chemical in order to reduce the penetration depth of the chemical compound into the façade material.
(b) Special safety precautions are necessary to protect the operator and the public.
   Acids will etch glass, attack stones with calcareous cement, aluminium windows and other metals, painted surfaces etc. So adequate scaffolding and screening are important.
(c) Great care must be taken to ensure that all of the applied chemicals are removed by copious water washing. Chemicals that are not completely washed off the pores may cause the appearance of efflorescence on the surface and salt decay will ensue.
(d) It is desirable to avoid high concentrations and long duration of action.

Some of the above-mentioned acids like hydrofluoric, oxalic and phosphoric produce insoluble salts, the remaining acids may react with some of the constituents of the material beneath the deposits and produce soluble salts.

Attention should therefore be drawn to the following disadvantages if acids are used that produce soluble salts.

(a) As already mentioned above, the acid will penetrate into the pores of the façade material, attack the surface layer beneath the deposit and thus cause the loosening, breaking up and disintegration of the surface layer.
A part of the surface layer will thus be washed away together with the dirt.

(b) The penetrated acid which is not completely washed out tends to dissolve iron minerals in stone and cause colour changes such as bleaching or patchy brown staining.

(c) Hydrochloric acid leaves chlorides behind which are very hygroscopic and which will increase the moistness of the wall.

Acids are preferred which produce insoluble salts with calcium.

Alkalis:
The procedure of cleaning with alkalis is to apply these substances on the dry soiled area, allow it to remain for a certain time and then wash it off completely with water.

Several repetitions of washing may be necessary.

An after-treatment with a neutralizing acid is not always preferred, it is a disputed case whether or not a neutralizing process is considered necessary.

Alkaline cleaners compete with water washing. The advantage of this method over water washing is the considerable reduction in the used quantities of water.

The action of alkalis is based on the saponification of fats or the neutralisation of fatty acids, so these substances are very suitable for the removal of oily or greasy dirt. Alkalis do not act upon calcareous substrates, but attack in slight measure some siliceous stone or glass.

The most used alkalis are based on caustic soda or potassium hydroxide. The main concern with caustic soda is that remnants of this soda left behind in porous stones may react with components from the atmosphere and produce sulphates and carbonates, which will cause salt crystallisation decay.

This process may be represented as follows :

$$2NaOH + SO_2 + \tfrac{1}{2}O_2 \rightarrow Na_2SO_4 + H_2O$$
$$2NaOH + CO_2 \rightarrow Na_2CO_3 + H_2O$$
or in the presence of a sulphate :
$$2NaOH + MgSO_4 \rightarrow Mg(OH)_2 + Na_2SO_4$$

The different forms of sodium salt hydrates such as $Na_2SO_4 \cdot 10H_2O$, $Na_2CO_3 \cdot 2H_2O$ etc. with their varying volumetric sizes, dependent on the relative humidity, will set up sufficient forces to rupture the surface.

The following unwanted effects must be taken into account when alkaline cleaners are used :

(a) It will etch glass, attack enamel, glazing and so on.
(b) It will harm the operators, if safety precautions are not taken.
(c) It may lodge in joints basicly in brick- or stonework and care should be taken to point open joints before hand and to wash the stonework thoroughly after cleaning.

Remaining chemical compounds :

- Complex compounds, the so-called chelate compounds, which are applied on the soiled area in the form of a paste. These compounds are mostly used to remove stains that consist of hardly soluble salts such as stains with high metal content.
It works by producing soluble compounds through reacting with the metal ions.
Gypsum crusts are also easily removed by the use of these compounds. The advantages are here that no harmful substances are left behind in the pores of the wall material and that washing-off is in some cases unnecessary.
Chelate compounds are suitable for cleaning entire walls as well as local stains.
- Chemical compounds to combat organic growth on walls such as mould, fungi, algae, moss.
Also to be mentioned are the chemical agents especially used to clean several kinds of stains, of frequent occurence are metal, oil and tar stains.
A large variety of above-mentioned chemical compounds are available and listed in the literature. A very large number of alternative treatments (recipes) also exists in the literature.

5.3.3 Mechanical cleaning

Dry blasting :
With this method abrasive particles are directed with
force onto the soiled area by means of a compressed air
jet. The cleaning action is accomplished by the impinge-
ment of the particles, by which a part of the surface
layer will also be pulverized and carried off together
with the dirt. The dislodging of the dirt deposits thus
takes place by the breaking up of the surface layer beneath the deposits to a depth of some millimeters.

The removal of a superficial dirt layer may take off a
millimeter of the stones but the removal of grime-impreg-
nated sandstone could require several millimeters. But
sometimes such a treatment could lead to destruction if
the crust is too thick. In that case a repair of the
surface is likely to be necessary.

The mean of transport is here compressed air, this has
the advantage that water penetration is not present.

The abrasive materials used are : metal or metalslag,
minerals such as silicon carbide, organic materials like
grinded pips or walnut bark etc.

Other variables to be determined are : the grain size
and the hardness of the abrasive, the flowing pressure of
the compressed air (generally 0.4 - 0.7 N/mm , in the
Netherlands also 10 - 15 N/mm   highest 20 - 60 N/mm).

The air jet can be screened in order to discharge the
blasting components directly together with the dirt. This
method has the following disadvantages :

   (a) Siliceous grit will produce dust that is a health
       hazard to the operator and the public.
   (b) A large volume of dust and a great deal of noise
       are generated, the dust will penetrate openings
       even in adjacent buildings and vehicles.
   (c) The possible protective hardened skin will be remo-
       ved and reveal the softer interior, this will make
       the cleaned surface more vulnerable.
   (d) When the surface hardness is not equal over the
       whole area the softer parts will be more eroded,
       giving an uneven textured appearance. The stone
       surface may thus be left in a damaged and unsightly
       condition.

5.3.4 Remaining mechanical cleaning methods

Wet blasting:
The abrasive particles are now blown at high pressure on
to the surface mixed with water.

This mixture of water and abrasive tends to be less
harsh than the dry abrasive.

The mean of transport is here water, this method is already mentioned in the section dealing with washing.

The same abrasive particles as used with dry blasting can also be applied with wet blasting.

Wet (grit) blasting does not produce dust as dry (grit) blasting, hence is favoured on health grounds.

For this reason silicious abrasive can also be applied with wet blasting. This method has also its disadvantages compared with dry-blasting namely :

(a) It generates water spray and run-off.
(b) It produces a considerable amount of slurry, which tends to make the method unpopular with operatives.

Some further dry methods to be mentioned are the removal by brushing, spin abrasion, scouring, grinding off, milling off. Brushing can be carried out with hand- or mechanical tools, almost no damaging is done to the material surface by brushing. Grinding and milling off are on the other hand rigorous methods suitable for the removal of hard, tough deposits.

Mechanical cleaning causes in most cases considerable damage to the material surface, which is not to be neglected. It is therefore advisable to weigh the disadvantaging effects of cleaning against the damaging effects caused by soiling before deciding to clean.

The use of iron tools or other metallic tools that can cause staining has to be avoided.

5.4 PROBLEMS INVOLVED IN THE TREATMENT OF SOILING OF BIOLOGICAL ORIGIN

Numerous tests have been carried out on different substrates to eliminate and prevent soiling due to microorganisms, using almost all the known biocides with various application methods (3, 7, 8, 10, 11). However, taking into account the fact that soiling persists in many places, a certain number of questions remain open.

In fact, experience shows that, although we dispose of polyvalent biocides such as "Javel water" (sodium hypochlorate), it is very difficult in practice to use products which have effects on all the types of microorganisms existing in soiling. Indeed, only products can be used harmless to substrate materials. For example Javel water would not be save on reïnforced concrete, because it produces chlorides.

Moreover, it has been proved that using an active product or one of excessive specificity can cause lack of balance in the ecological association treated and lead to proliferation of a certain type of microorganism to the detriment of others (11).

Biocide products are not generally lasting. Since
their effect is of limited duration, periodic renewal is
necessary. In this case, there can be fear of certain
microorganisms undergoing genetic mutation which will give
them resistance to the products used. A comparable
phenomenon occurred in agronomy a few years ago when it
was found that D.D.T. treatment against certain insects
was no longer effective, whereas several years before it
has been remarkbly active.

It is equally interesting to note that the use of such
biocides is not always sufficient to eliminate soiling and
that before undertaking restoration, it is necessary to
carry out complete cleaning to properly eliminate microorganisms (12).

In certain types of products - paints in particular -
algicides and fungicides can be incoporated in the formulation (4,6), but there seems doubt that these
give complete guarantee as to the result obtained.
Indeed, soiling due to microorganisms perhaps takes longer
to appear, but it is nevertheless not completely
prevented.

At the present time, the products most used are mainly
quarternary ammonium salts (1, 2, 12). Other types of
active products,   for example copper oxiquinolate (9),
chromium trioxide and copper sulphate can also be used,
but they often imply parasitic colouring of materials.

Moreover, more complex physical methods using certain
types of radiation would seem to offer wide polyvalency
(5) provided they can be used easily and economically to
eliminate soiling due to microorganisms.

In fact, in all the cases where curative treatment is
necessary, it seems important to carry out preliminary
tests using sufficiently reliable laboratory techniques,
properly representative of practical conditions (1, 2).

A final appraisal shows that, at the present time,
there is no cleaning method sufficiently lasting.

There is obviously a need to perfect curative treatment
but endeavour must particularly aim at preventive methods
to act on this type of soiling.

Indeed, it has been noted in this report that water is
an important factor in the growth of microorganisms of
plant origin. Spores transported by the wind will remain
in this vegetative state as long as the conditions favourable to their growth and breeding are not brought together. Consequently, we can consider that one of the
surest means of preventing the occurrence of soiling due
to these microorganisms could consist of eliminating any
possibility of water absorption or stagnation on façades.

Knowing that the force exerted by microorganisms to
absorb water is not infinite, it seems that the size of
pores in material must also be taken into account because
if they are small enough, the capillary tension forces
will be sufficient to make it impossible for the micro-

133

organisms present to use any water contained in these pores.

.

References

(1) Grant C. and Bravery A.F., (1981) Laboratory evalua-
tions of algicidal biocides for use on constructional materials. 1. An assessment of some current test
methods, International Biodeterioration Bulletin,
17, 4, 113-123.
(2) Grant C. and Bravery A.F., (1981) Laboratory evalua-
tions of algicidal biocides for use on constructional materials. 2. Use of the vermiculite bed
technique to evaluate a quaternary ammonium biocide;
International Biodeterioration Bulletin 17, 4,
125-131.
(3) Higgins D.D., (1982) Removal of stains and growths
from concrete, Cement and Concrete association,
Appearance matters, 5, 11 p.
(4) Hoffman E., Hill R.K., and Barned J.R. ; (1976)
Fungus resistant paintsfor the humid tropics J. Oil
Colour Chem. Assoc., 59, 2, 62-68.
(5) MacCarty S., (1981) Cooling tower microorganisms
controled by ultraviolet contact ; Heat. Piping
Air Cond., 53, 10, 63-65.
(6) Rabate H., (1975) La peinture dans la construction.
Environnement, pollution et souillures biologiques
des revetements par peintures et préparations
assimilées et par feuilles de matières plastiques;
Revue Technique du Bâtiment et de la Construction
Industrialisée, 22, 47, 89-97.
(7) Rechenberg W., (1972) Verhinderung und Beseitigung
von Algen- und anderem Bewuchs auf Beton, Beton,
22, 6, 249-251.
(8) Richardson B.A., (1975) Control of moss, lichen and
algae on stone ; the conservation of stone, Pro-
ceedings of the International Symposium, Bologne
225-231.
(9) Singh S.M., and George J., (1965) Effect of some algicides on the strength development of cement paints;
Paintindia Annual, April 129 and 130.
(10) (1963) The control of lichens, moulds and similar
growths on building materials ; Building Research
Station Digest, reprinted 1968, first series 47, 1-4.
(11) Trotet G., Dupuy P. and Grossin F., (1975) Traitement
chimique de la piste de la base de Landivisiau ;
Revue Générale des Routes et Aérodromes, 49,
509, 74-77
(12) Whitheley P. and Bravery A.F., (1982) Masonry paints
and cleaning methods for walls affected by organic
growth ; J. Oil Colour /chem. Ass.
1, 25-27.

Chapter Six

Recommendations for design of building façades

6.1 INTRODUCTION

Architects have to consider many aspects when designing
building façades which will influence the final result
both aesthetically and technically.
 These aspects include :

- The client's brief which will establish standards of
  quality and cost limits.
- The location, that is such matters as town planning
  requirements, climatic considerations and admissible
  loads for the type of ground, etc.
- Technical aspects, that is daylight and insulation
  standards and the choice of suitable materials.

In the past many buildings have been spoiled by the
failure of designers to pay sufficient attention to their
long-term appearance. Controlling changes in appearance
throughout the life of a building demands particular
attention to their situation and choice of suitable
materials and details.

6.2 THE SITUATION (LOCATION)

A number of fixed data governs any location, for instance:

- Composition of the air
- Climate
- Surrounding development

6.2.1 Composition of the air

The polluting elements in the air are of a certain colour
and the heavier the pollution, the more quickly and
distinctly a building becomes soiled. In most urban
situations the predominant colour is black or grey but in
some middle eastern areas for instance, the colour of the
local desert sand becomes significant. The architect
therefore can familiarize himself with the possible
effects soiling will cause on the planned building. Such
an exercise would consist of the following investigations :

(a) The state of the surrounding buildings.
(b) Investigation of the local sources of pollution,
    e.g. factory or busy road network or vegetation.
(c) Published information e.g. black smoke or SO_2
    levels.

It is known that the air closer to earth is more polluted
than at higher altitudes. Pollution and the stratification thereof can be reason enough for the designer to take
such design measures as :

- Providing different materials for the most difficult
  parts of the façade e.g. easily cleaned surfaces at
  ground floor level.
- Providing bold details to divide the façade horizontally.
- Using darker colours in dirty areas.

6.2.2 Climate

In dry periods the façade that faces the prevailing wind
direction will soil more quickly and heavily. The combined effect of the prevailing wind and the stratification
of pollution prevents the even distribution of dirt, but
the most visible problems of soiling are caused by the
movement of water on the façade. In the western part of
Europe the wind bringing the most rain comes from the
south-west. The façades facing this direction are therefore subject to a great deal of water which cleanses them
more or less effectively depending on the absorptivity of
the façade material (Chapter 3.4.).
    Some of this water is carried by the wind circulating
over and around the building and becomes concentrated on
parts of the façade where the wind abruptly changes its
direction, such as the upper parts of the building and at
corners.

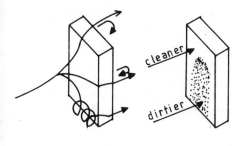
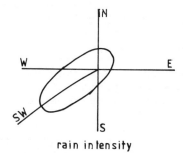

rain intensity

fig. 6.2.2. 1.

The orientation of a building also affects the amount of sunlight which reaches each façade. North facing façades will be colder and damper than south facing façades. A damp material is normally darker than a dry one while also a damp surface is an ideal source of nutrition for living organisms such as algae. North facing façades require particular attention to detail when constructed in porous materials. Special attention must also be given to buildings or parts thereof which receive little or no rain because they are situated in the lee of the wind (i.e. in the rain shadow). This may be because of the surrounding buildings or afforestation but also can be due to the effect of the detailing of the façade. Moreover if this situation is coupled with rain running down the façade, one of the least acceptable forms of soiling is created. It is therefore essential to protect that part of a façade lying in the rain shadow from rain streaming down it.

6.2.3 Surrounding development

Surrounding development can dramatically change the wind and rain direction and amount and can for instance shade the sun from the south façade. The points considered under 6.2.2 should always be studied in relation to the particular site and any adjacent buildings or natural features.

6.3 FACADE MATERIAL

The architect can give a certain significance to a façade by the use of specific materials. Materials like brick cement rendering, concrete, plastics and even paint may form the outermost layer of the façade. The choice of the façade materials is dependent on costs, technical demands and the owner's wishes but is also affected by the designer's desired expression of the façade. For example natural stone and bronze can give an impression of luxury and wealth to a building and for that reason these materials have often been used for banks, civic buildings, etc. while plastics and bright colours have been used more often for factories and commercial buildings. Façade materials vary in their ability to resist the influence of the weathering effects, but every material and so every façade alters in appearance after long exposure to wind and rain.

In chapter 3.4. "rain run off" it is explained that dependent on the quantity of driving rain within a given time, water run off only occurs if the absorption coefficient of the façade material is lower than a certain value. So the denser the material the quicker run off occurs.

It also explains how low the absorption coefficient has to be (dense material) for the water film to reach the ground level.

Table 1 : absorption coefficient $g/m\ s^{\frac{1}{2}}$ for some materials.

rendering	40-250
hard-burnt brick	125
limestone	10-100
concrete	10- 40
metal, glass, plastic	0

Roughly one can divide façade materials in three groups
- Materials with a high absorption coefficient (e.g. rendering, brick and some types of limestone).
- Materials with a medium to low absorption coefficient (e.g. concrete and many types of stone).
- Materials with no absorption coefficient (e.g. plastics, glass and aluminium).

6.3.1.1 Materials with a high absorption coefficient
Most of the water hitting the surface will be absorbed by this type of material. The façade normally remains damp for a longer period than a façade with denser materials and soiling particles adhere easily to this type of material. Provided that major differences in rainload on adjacent parts of façades are avoided soiling will often be evently distributed over such surfaces.

6.3.1.2 Materials with a medium to low absorption coefficient
Concrete usually has a medium to low absorption coefficient but there are many different types of concrete and architects must choose casting methods and mixes suitable for the various situations on a building façade. In general good concrete quality has a low water cement ratio and has been well compacted. The outer millimetre or two forming the skin of the façade cast against a mould surface can be very variable and can be affected in its early lifetime by the different curing regimes. It is often easiest for the long term appearance of the concrete if this skin is removed. The most obvious effects on materials such as concrete and stones with similar medium to low absorption coefficient is described in 3.4 from which it can be seen that with these materials special care has to be taken to ensure that water is not allowed to run unchecked down surfaces unless there is enough to wash them completely.

6.3.1.3 Materials with no absorption coefficient
This type of materials gives a very quick removal of water
from the façade. Typically water runs over such surfaces
in discrete streams rather than as a continous film. In
polluted areas this can produce dirty streaks and may
periodically require cleaning. In the design process of a
building special attention should be given to the effect
of the run-off water from these dense materials onto
underlying materials with different absorption coefficients and also any harmful effects which dirt may have
on the material itself.

6.3.2 Changes in the skin caused by the environment

See chapter 2.3.

6.3.3 Inter-relation of the materials used in the façade

See chapter 2.4.

6.4 THE PROFILING OF THE FORM AND THE
DETAILING OF BUILDINGS

As suggested in 6.3.1.2 water can be used to keep surfaces
clean if the quantity of water is sufficient in relation
to the absorption coefficient of the material, while 6.2.2
explains why this does not naturally happen on all
building façades. On the lee side of buildings special
means would be required to collect enough water.
    In "Sallisures de façades" Carrie and Morel suggested
that panels or whole façades can be sloped in order to
increase the amount of water on the surface, and where this
is suitable for the architectural intentions of the
building it would be satisfactory. However, not all buildings can be designed in this way which must mean that
many buildings cannot be naturally washed clean and must
retain a layer of dirt on their surfaces.
    In this situation - normally on the lee sides of buil-
dings particular care must be taken to choose materials
which will not be spoiled physically or visually by the
dirt and also to ensure that the soiling is allowed to
build up evenly without disturbance from uncontrolled
water.

6.4.1 Guidelines

A façade can become spoiled through the pollution deposited on it and by the rain flowing over it. The following guidelines can aid the design stage.

 Control the flow of rain over the façade so that:
 (a) Water is not allowed to run unchecked down surfaces, unless there is enogh to wash them completely.
 (b) No water remains on the façade.
 (c) Water is not permitted to flow onto vertical surfaces from any horizontal surface.
 (d) No flows of water occur over a surface in the rain shadow.
 (e) Water containing a high concentration of pollution is quickly drained away.
These guidelines should be applied to all parts of buildings. We give below two examples : a simple horizontal strip or string course and a large blank façade.

6.4.2 Example 1 : Horizontal strip or string course

surface A, A1   (horizontal, oblique) upwards alignment

surface B,   vertical alignment

surface C, C1   (horizontal, oblique) downwards alignment

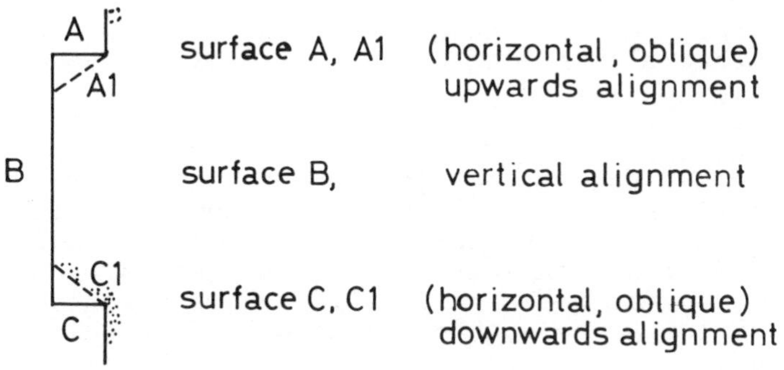

fig. 6.4.2. 1.

The difference in alignment of surfaces A, B and C will cause different concentrations of adhesive and accumulated dirt. In dry periods the most soiling will be found on surface A, though B and C can be soiled by the finer particles in the atmosphere. Surface A also catches most rain. At the top of a building the slope of the rainflow can be nearly horizontal in which case the whole detail would be likely to remain clean, but at lower levels the rain falls obliquely on to the façade or even vertically.
 Thus surface A related to the slope of the rainfall gets much more water than surface B. The absorption and saturation capacity of surface A is quickly reached and

water flows on to surface B. If the concentration of
pollution in this water is high it may cause dirty streaks
on surface B. Usually, however, it will flow irregularly
over the edge only partially washing surface B. On smooth
surfaces with medium to low absorption coefficients such
soiling of horizontal ribs is clearly visible.

If surface A has a high or medium absorption coefficient and is allowed to remain damp it will support the
growth of algae and moss.

Water from such a surface should be drained off directly without it coming into contact with the other aligned
concrete surfaces. If this is not possible such ledges
should be inclined as surface Al. Surface C must have a
drip groove or other feature to prevent water from flowing
onto the area in the rain shadow. If it has no groove or,
worse still, is aligned as Cl it will introduce localized
flows of water onto the vertical surfaces below.

6.4.3 Example 2 : Façade without windows

We will consider in turn the likely performance of façades
faced with a) highly absorbent materials, b) medium-low
absorbent materials and, c) non-absorbent materials, when
they are facing the prevailing wind and when they are on
the lee side of a building.

The most difficult circumstances occur where there are
few windows or other elements of design which can be used
to control the flow of water.

(a) High absorbancy
Beyer's study suggest that façades of high absorbancy materials will rarely receive enough rain for
a flow of water to develop. However the façade will
still be getting much wetter at the top and around
the edges, so materials which will change colour or
produce effloresence with constant wetting should
be avoided.
The top edge of such materials must always be protected by a coping or a flashing designed to prevent water from getting from the horizontal surfaces (roof or top of parapet) onto the vertical
face. This should be designed to direct water onto
the roof or if it has to slope outwards, it must
have a very bold overhang as small projections at
high level have been shown to cause problems with
upward-driven rain. Special attention must be paid
to the jointing of such cappings as any failure
will cause unsightly streaks on the façade.
Sheltered façades do not get as wet as those on the
exposed sides of buildings but they stay damp
longer because there is less sun and wind to dry
them. In these circumstances mould and algae may

142

develop on materials with high absorbancy coéffi-
cients.

(b) Medium to low absorbancy
It is clear from Beyer's work and from observation
of existing buildings that partial washing is
particularly associated with materials with medium
to low absorbancy. On large façades any heavy rain
will quickly saturate the top part of façades and
start to flow down. When this flow reaches the
lower level where rain hitting the façade is insufficient to saturate the material it will be
absorbed and will deposit any dirt that it has carried down from higher on the wall. The irregular
dirt line thus formed is characteristic of such
materials as concrete and many porous stones in
such situations.
This phenomenon can be controlled only by chan-
neling the water or by preventing the flow from
developing. Horizontal string courses or simular
details bold enough to throw water off the façade
will serve several functions at once. They will reduce the amount of water on the surface, reduce the
differences between panels at different levels on
the façade, make the change from washed to unwashed
into a gradation instead of a clearly visible line
and by producing interest and shadows will make any
changes less noticeable.

(c) Non-absorbancy
Rainwater hitting façades made of completely impervious materials must, unless it evaporates, flow
right down to the bottom. If it can be induced to
flow in an even film it should wash the whole
façade clean. However, most such materials are
hydrophobic and water running on them tends to form
discrete streams. Such materials can therefore
become noticeably streaked. Whilst many building
materials can carry a certain amount of dirt
without much harm to their appearance or their performance, most impervious materials look best when
they are clean and bright. Some such materials can
be harmed if dirt is permitted to accumulate on
their surfaces. In heavily polluted locations
façades made of such materials may need to be
cleaned more frequently than more porous but less
vulnerable materials.

6.4.4 Facades with apertures

The two previous examples mentioned had to do with
different concentrations of adhesive and accumulated dirt
caused by a difference in alignment and also the likely
performance of facades without windows. Both examples
divided the materials up into units according to their
percentage of absorption.

In facades with apertures we are confronted with
differences in material such as glass and brick. One
exception is the application of 'structural glazing
systems' which almost only consist of glass, but these are
rare.

What is important in the application of different
materials in the facade is the difference in 'absorbancy',
causing discolouration and even algae growing under the
glass surfaces.

In western Europe, this happens especially on facades
facing North or East. Facades facing south and west
receive a much larger amount of rain and therefore will be
washed clean more easily. Dependent on the absorbancy of
the facade material, dirt which has been carried down
will be deposited at the end of the rainflow on the
facade, where it will cause an irregular dirt line. Glass
surfaces lying in the same surface with a more absorbant
material in the facade, must have a provision on the
bottom of the glass surface designed to guide the
concentrated waterflow off the facade.

By mounting the glass surface more to the innerside of
the facade surface, the latter acquires more plasticity.

But as a result of the form special attention is necessary for medium to high absorbance materials.

Correct detailing is therefore necessary to incorporate the effects of soiling. A water barrier (water hole) is necessary in the upper part of the window niche. The wind will then drive the water via the water hole to the side of the niche. If no precautions are taken, the water will stream down one of the sides walls of the niche. The soiled patterns which can develop are to be prevented by :

- ending the water hole approx. 30 mm from the side wall
- continuing the water hole in the side walls

6.4.5 Ribs

Ribs can be used by the architect to give a special expression to a facade but also to control flowing along the facade. In principle ribs can be employed in every conceivable direction. In general the following differentiations can be drawn :

Horizontally placed ribs

These can be used to protect the underlying surface. They can also be used on the surfaces subjected to little rain or large surfaces such as side facades as they drain water away from the facade. A water barrier in the downwards aligned surface is necessary.

Vertically placed ribs

Vertically placed ribs are used by the architect in zoning
rainwater down the facade. When a single rib is used rain
will mainly fall onto its outer edges because of the
curving influence of the wind over the rib. The water
flows off the outer edges thereby cleansing them if there
are used less porous materials. Deposited pollution is
concentrated mostly in the grooves. Their darker colour
accentuates the vertical profiling. The rib must not be
too wide otherwise a soiled pattern develops in the
middle area of the rib's upper surface.

Arbitrarily placed ribs

When ribs are arbitrarily placed, highly polluted water
gets ejected from the end of the rib and falls onto the
facade. Therefore the ribs have to be very short to avoid
such unaestetic soiling. The ribs can also be used in
this way to give the surface of the facade a certain
roughness. The arbitrarily placing of many short ribs
causes a greater regularity of the soiling pattern of the
facade. The effect of the shadow cast gives the facade
vitality. A facade can also be designed with upright
ribs, closely placed, that touch each other atcertain
corners (e.g. The Bijenkorf, a department store in
Rotterdam). Soiling patterns are not easy to predict. The
more horizontal the rib, the greater is the deflection of
the flow of water from   the rib's alignment.

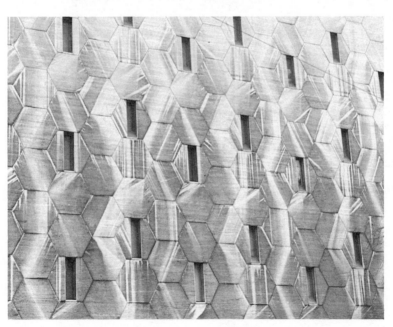

RAINSHADOW

The detailing of the the form of the aluminium entrance
prevent rain run-off all the parts of the entrance.   In
the rain shadow area soot included chloride products
attack the aluminium.

location	mediumsize city
date	1968
orientation	north east
material	prefab concrete
special features	no extern. treatment

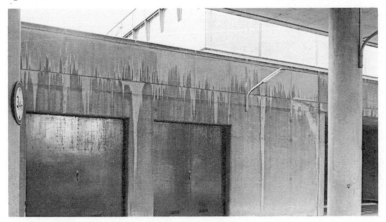

This part of the building is situated under the rain
shadowed area of another building.   Here is a good example
where we can see why a rainshadow area from a more porous
material needs to be protected from falling water.   A cap
piece sticking out of the facade can provide a water
barrier that will allow the facade to evenly discolor.

ORNAMENTAL FORMS

location	large city
date	1965
orientation	east
material	in situ concrete
special features	no extra treatment

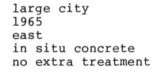

present situation
top-side

alternative

The balcony acts as a water catcher and holds the under-laying parapet in the rainshadow area. At certain places, the facade is washed clean, leaving a contrast between the clean and soiled areas of the facade. The illustrated change in the cap piece detail may prevent water from streaming down the facade.

CONCENTRATED WATERFLOW ACROSS THE FACADE

location   medium size city
date   1970
orientation   east
material   prefab concrete, washed
                            out and afterwards grit
                            blasted

special features

deviation of rain by non
constant wind pressure

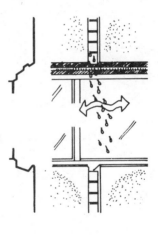

present situation

Facade on which little water falls. The water that
collects the gutters under the windows, falls to the
ground, via a vertical gutter, that is designed as a
'cloister'groove - over the parapet. On the bottom of the
parapet the water falls directly down. Under the influence of the wind the falling water causes this typical
soiled pattern on the facade.

149

CONCENTRATED WATERFLOW ACROSS THE FACADE

location	large city
date	app 1968
orientation	north east
material	prefab, no extra treat-ment

special feature

Facade where little water falls. The water that collects
in the gutters under the windows falls down via the
working joint between the elements. The joint filling is
sunken in the element where by a cloister grove arises on
the bottom of the parapet. The water falls directly down
under the influence of the wind and causes this typical
soiled pattern. At these places the cement is dissolving.

HORIZONTAL RIBS PROTECTING THE UNDERLYING SURFACES

location	large city
date	about 1985
orientation	north east
material	prefab concrete, parti- ally washed out

The design is intended to let the rainwater fall from the
facade as fast as possible. Under the ledge that stick
from the facade, a rain shadow area arises in which
soiling leads to a darker tint. Visually it gives the
impression of a normal shadow area.

influence caused by
interruption of the
horizontal profile

DENSE MATERIALS

location   large city
date   unknown
orientation   east
material   metal

A renovated building cladded by a metal façade shows that
apertures lead to a partial greater rain-offer on the
façade which leads to partial washing. Regular washing of
this type of façade is necessary.

In the design process of
a building special at-
tention should be given
to the effect of the
run-off water from dense
materials onto under-
lying materials with
different absorption
coefficient.

COMBINATION OF MATERIALS

location   medium size city
date   1978
orientation   south east
material   prefab concrete, washed
                            out
special features   metal roof (capping)

The front side of the metal roof and the facade lie in one
line. Placing the roof slightly behind the facade and
letting the rainwater fall in a gutter can prevent the
facade from being soiled.

COMBINATION OF MATERIALS

location   medium size city
date   1971
orientation   east
material   prefab concrete, washed
                            out

special features

This building has a horizontal structure. Thus the putty
between the prefab elements must be inconspicious. The
putty has dispersed over the concrete surface. Because
the putty continued to be sticky, dirt attracted to it
easily. Cleaning the concrete surface and replacing the
putty will lead to an improvement.

Changing the skin of
concrete in mosaic tiles
does not change the
soiling configuration

154

APPENDIX 1

The sandblast test as a method to judge the properties of
the surface.

Introduction

When building materials have been exposed to the outside
weather conditions for many years, one might wish to know
how much and to what depth this material has been
affected.
  By measuring the abrasion resistance (or erosion velocity) on different depths from the outside material layer,
one obtains a reliable indication of the surface condition. Besides testing old weathered samples, it is also
possible to make a judgement of rather new surfaces. This
might for instance be of interest with concrete of the
same composition but with a different appearance. This
can be due to several circumstances such as temperature
during hardening, method and time of vibration, conditions
after demoulding etc. Then the question may arise as to
how far this can influence the quality of the toplayer of
the concrete. Here also the abrasion resistance oan give
us sufficient information. Abrasion resistance is
measured by using the "soft" sandblast method.

Test-samples

In order to measure the abrasion resistance at different
depths, at least 7 equal test samples must be selected.
The samples must have an area of 8 x 8 cm. Counting from
the toplayer of the material layers of 0-0.5-1-2-3-5 and
9 mm are removed.
This is done by polishing the test-samples until the
desired depth is reached. Then the samples are stored at
20°C and 50 % R.H. until they will have reached a constant
weight.

Method

The "soft" sandblast test is carried out using the sandblast apparatus of Vogel and Schemmann, such as described
in NEN 7000.
  The principle of the method is based on the loss of
weight of the sample, after   the surface has been
sandblasted during a certain time and at a certain
pressure. Then this loss of weight is compared with the
loss of weight of standardized float glass that has been
sandblasted using exactly the same test conditions.
  In this way it is possible to compare the results with
other "soft" sandblast results obtained at an earlier

stage or elsewhere, when the same time and pressure were used during the tests.

The sand used in the test is standard sand which consists of clean quartz sand containing not more than 2% of particles larger than 1.2 mm and not more than 2% of particles smaller than 0.42 mm.

The area to be tested is a circle with a diameter of 60 mm. During the test the sample is rotated round its centre to produce an as uniform a treatment of the surface as possible.

The time and the pressure during the test are established at 30 seconds with a pressure of 2 atm. In case of very weathered surfaces the pressure can be lowered to 1 atm., but then one must realize that no comparisons can be made with test carried out at 2 atm. pressure.

Results

For each depth examined the loss of weight in the test material and in the standarized glass can be compared using the formula :

$\dfrac{\Delta \ \text{material}}{\Delta \ \text{glass}}$

APPENDIX 2

Test results of the "soft" sandblasting method

Introduction

Two practical applications of the soft sandblast method
are given : firstly the judgement of the weathering of an
old concrete block ; and secondly the judgement of a
concrete surface with light and dark locations next to
each other.

Test samples

The weathered concrete block was a 15 year old
prefabricated block that had been lying horizontally in a
sandbed during all this time, with only the top half exposed to the atmosphere (see fig. A2-1).
   Due to its horizontal position the upperhalf had had a
very large amount of soiling components deposited on it.
   The part stuck in the mainly wet sand, on the contrary
had had very ideal hardening conditions during all those
years. No frost damage or other damages were observed.

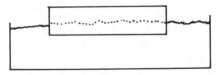

   fig. A2-1

   The second sample consisted of a large prefabricated
concrete element with light and dark coloured locations on
it. The differently coloured stains became visible rather
soon after demoulding the concrete element and did not
disappear for some months afterwards.

Method

The concrete block was first divided into two halves by
sawing it lengthwise to separate the part that had been
exposed to the open air from the part that had remained in
the sand.
   From the resulting halves, test samples were sawn of
8 x 8 cm, a total of 7 samples of each half.

The first test sample of each half block remained in its original condition. From the second two test samples a layer of 0.5 mm was removed by polishing, and from the next samples, layers of 1, 2, 3, 5 and 9 mm respectively, were removed in the same way. The 14 samples were then examined by measuring the abrasion resistance using the method of soft sand blasting. Each test sample was sandblasted for 30 seconds with a pressure of 2 atmosphere. Also three standard glass samples were treated in this way. After calculating every loss of weight it was possible for each test sample to make the equation

$$\frac{\text{loss of weight test sample}}{\text{loss of weight glass}} \quad \text{(see fig. A2-2)}$$

A somewhat different procedure was followed in the examination of the element with the colour differences, as these were thought to be only superficial differences.

Accordingly, a dark and a light coloured sample were first sandblasted with a pressure of 0.2 atm. to a depth of 0.05 mm. (This depth was calculated on the basis of the loss of weight and the specific mass of concrete). The time needed to reach this depth was noted. Then the test samples were sandblasted further with the same pressure of 0.2 atm. until a depth of 0.15 mm was reached and again the time was noted.

The next step was to sandblast to a depth of 0.3 mm but with a pressure of 1 atm; then to a depth of 0.9 mm with a pressure of 2 atm. and finally to a depth of 2.7 mm with 3 atm. pressure. For each step the time was noted.

To make the comparison with standard glass, 4 glass plates were sandblasted with the same four pressures as used with the test samples. The results are shown in fig. A2-3.

Using this graph, one can calculate at each pressure the depth of glass removed in the time taken for removal of a certain depth of concrete from the light and dark test samples. So at 0.2 atm. to reach 0.05 and 0.15 mm - at 1 atm. to 0.3 mm - at 2 atm. to 0.9 mm and at 3 atm. to 2.7 mm. In this way one can compare the different depths that were reached with the test samples and the glass during the same times of sandblasting. Results are shown in fig. A2-4.

Results

Fig. A2-2 shows the different erosion velocities of the exposed half and the half in the sandbed. The following remarks can be made :

(a) A very high difference of almost a factor of 5 in erosion velocity between the exterior parts of the weathered side and the well-hardened side.

(b) At a depth of 5 mm there is after 15 years of exposure no difference from the well-hardened concrete block.

(c) The main weathering is limited to the first two mm.

Fig. A2-4 shows the abrasion resistance of the light and dark coloured locations on the concrete element, expressed as the mm loss in depth against the same loss in depth of glass during the same time.

It will be obvious that the line of glass is drawn at an angle of exactly 45°.

Notice that the light part in the first 0.15 mm is softer than the standard glass, for the gradient of the line is deviatiny. Then the light part becomes equal to the glass.

The dark part, on the other hand, is harder than the glas until a depth of 0.3 mm and then its hardness equals that of the glass.

The cause of the lightness and softness of some parts versus the darkness and the hardness of the other parts is to be found in the fact that the latter had a lower water/cement ratio than the former - a fact which itself is probably due to an unequal loss of surface water and/or to local segregation.

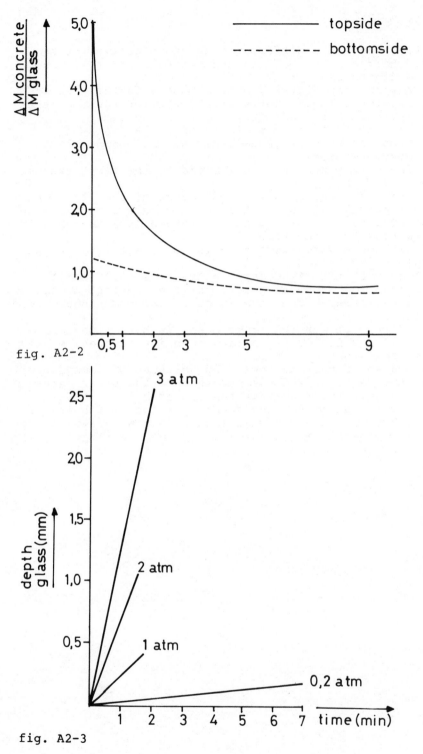

fig. A2-2

fig. A2-3

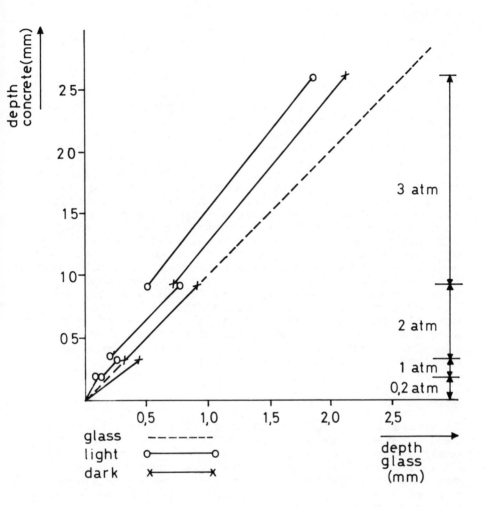

fig. A2-4

A method for determining the soiling capacity of air.

Introduction

To characterize the air quality of a certain environment
in order to measure the soiling capacity, there exist
several methods. For one or more reasons those methods
fail for measuring the real soiling capacity of façade
surfaces. Not the mass of the particles is the important
factor for the soiling of the façade, but the difference
in colour and reflection. Further, by suction a selection
takes place of the particles in the air. In the microclimate of the direct environment of the facade the
concentration and the size distribution of particles can
strongly differ from the average air sample that is
obtained with a suction method. It seems to be more
useful to "measure" the façade material itself. Therefore
in this report a method is introduced by which the
measuring of the façade material is done with the aid of
the reflectance. The decrease of the reflectance of
several concretes made from different types of cement can
give a more reliable prediction of the soiling capacity of
the environment. Furthermore 4 vertical surfaces are
observed in different orientations.

Method

To determine the soiling capacity of the air, there is
used a method with which the reflectance is measured of
small concrete façades.   (See fig. A3-1.)
   In the open field are placed four concrete cubes and
one cube of polymethylacrylate (PMMA) with the sizes of 25
x 25 x 25 cm . The concrete cubes have all the following
standard composition :

```
 450   kg cement
1440   kg gravel 3-8mm
 580   kg sand
 247,5 kg water
```
                  so the w.c.f. = 0,55

The only difference was the type of cement :

   1. Portland-A cement   ENCI
   2. Blastfurnace slag-A cement Cemy
   3. Flyash cement   ENCI
   4. White-cement   Dijckerhof

The cubes are fixed on standards at a height of 150 cm.

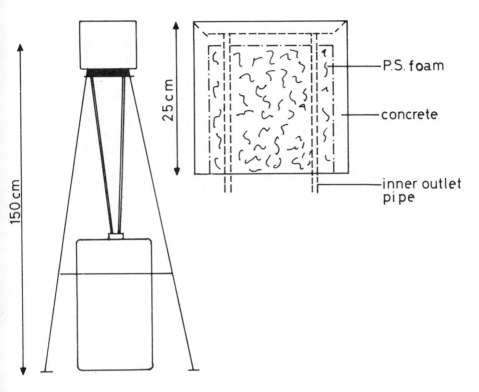

fig. A3-1.

The concrete cubes are so designed that the rain and the deposited dirt falling on the horizontal part are carried off through the inside of the cube. Therefore this amount of rain and dirt are prevented to run over the vertical sides.

This water and dirt are collected in containers.

Each concrete cube is build up out of four vertical and one horizontal plates, with a thickness of 3 cm, while the inner part consists of polystyrenefoam, surrounding the inner outlet-pipes.

The PMMA cube is only used as container for the rain and deposited dirt for there is no topside. The rainwater and dirt that fall into the cube are also collected in a container.

When sun radiation strikes the surface, the temperature of the surface is also influenced by the presence of the foaminside. So the texture, porosity, temperature and moisturecontent of the outer skin does not differ to largely from a real concrete slab. Of course the micro climat around the concrete will be different from that of a real building. Therefore it will not be possible to predict only from the cube the soiling of the concrete of a façade, but it gives an indication of the possible soiling capacity of the air in a certain region.

With the aid of those concrete cubes, the soiling capacity of the air will be determined by measuring the change in reflectance of the vertical sides. The reflectance is measured after regular time intervals on the four sides of the cube, of which each gives the soiling capacity of the corresponding wind direction. Those winddirections are south-west, south-east, north-west and north-east.

The reflectance of each side is expressed as the average reflectance of four measures, for the reflectance apparature measures a square of 10 x 10 cm . The reflectance itself is measured with a small portable apparatus, that is adjusted with the aid of two photographical test-cards (darkgrey and white) representing reflections of 18 resp. 90 %.

APPENDIX 4

Test results of the reflectance measurements

In order to test the possibilities of the apparatus, cubes
were placed at three places in Holland with three
different environmental conditions. At each place a
similar series of four cubes was installed, three of them
consisting of different types of cement and the fourth
consisting of PMMA.
 The environmental conditions at the three chosen sites
be characterised as follows :

- Velsen   Situated at about 100 m from the sea
                   and a few hundred meters south of a
                   large blast furnace.
- Amsterdam   Between light industry and several
                   small gardens in the immediate vi-
                   cinity.
- Eindhoven   In the middle of an urban district
                   without industry.

At these three places the following data are being
collected :

- dust concentration
- concentrations of SO_2, NO_2 and some heavy metals
- pH of rainwater
- metereological data

The results of the decline of the reflectance during 20
months of exposure at three different locations are shown
in fig. A4-1.
 Also shown are the amounts of calcium (Ca^{2+}) dissolved
by the rain that had fallen on top of the cubes.
 The quantities are corrected for the amount of calcium
that had fallen as deposited dirt.
 Tables 1 through 3 present the pH, the amounts of
calcium (Ca^{2+}), the amounts of deposited material, and the
amounts of rainwater collected in the container below each
cube, for each site.

Location Velsen

	pH		
time (months)	3	17	20
PC	6.7	6.35	6.6
BFS	6.75	6.4	6.4
Fly	6.7	6.3	6.5
White	6.7	6.2	6.0
PMMA	4.0	3.6	3.8

	tot. Ca (mg)			
time (months)	3	17	20	tot.
PC	216.5	1032	262	1510.5
BFS	322	1220	187	1729
Fly	237	1137	260	1634
White	190	1144	263	1597

	tot. deposited (mg)			
time (months)	3	17	20	tot.
PC	1017	5457	891	7365
BFS	1003	5454	457	6914
Fly	806	5952	599	7357
White	757	5691	816	7264
PMMA	1938	7960	1323	11221

amount of water collected in the container (1)

time (months)	3	17	20	tot.
PC	9.025	46.700	12.200	67.925
BFS	8.105	45.300	9.600	63.005
Fly	10.680	51.930	13.970	76.580
White	8.525	51.300	11.505	71.330
PMMA	13.435	69.700	16.470	99.605

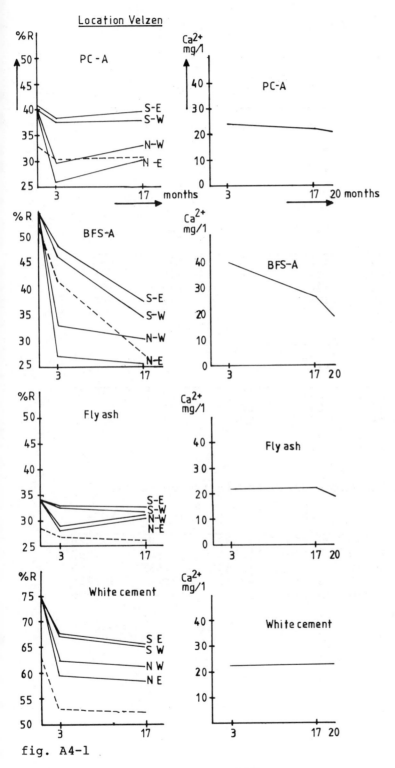

fig. A4-1

167

Location Eindhoven

	pH			
time (months)	4	13,5	17	20
PC	6.65	7.3	6.8	7.2
BFS	6.7	7.3	7.0	6.5
Fly	6.5	7.1	6.7	6.6
White	6.5	7.0	7.0	7.1
PMMA	4.35	4.8	4.7	3.9

	tot. Ca^{2+} (mg)				
time (months)	4	13,5	17	20	tot.
PC	83	431	181	147	842
BFS	143	374	166	95	778
Fly	90	494	188	129	901
White	95	476	178	151	900

	tot. deposited (mg)				
time (months)	4	13.5	17	20	tot.
PC	23.5	27	107	99	256.5
BFS	47	83	106	91	327
Fly	36.5	40	77	74	227.5
White	25.5	42	97.5	58	223
PMMA	123	291	157	145	719

	Amount of water collected in the container (1)				
time (months)	4	13,5	17	20	tot.
PC	4.030	26.950	13.300	7.200	51.480
BFS	3.660	23.360	11.875	4.900	43.795
Fly	5.970	31.240	14.045	6.635	57.890
White	6.390	32.180	14.380	7.265	60.215
PMMA	10.090	45.730	18.960	10.400	85.180

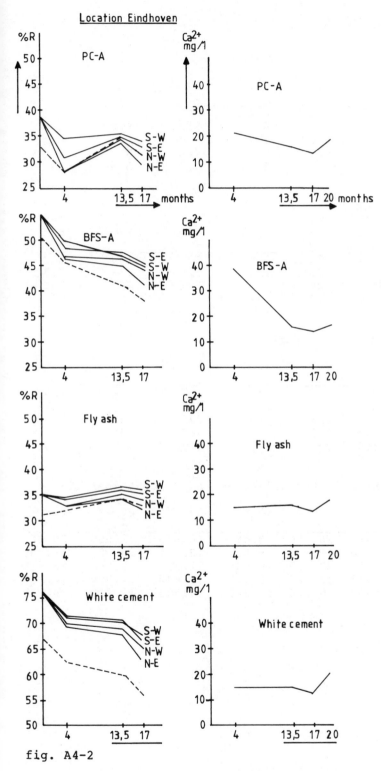

fig. A4-2

Location Amsterdam

pH

time (months)	3	9.5	13.5	17	22
PC	6.45	5.8	6.1	7.1	6.0
BFS	6.6	6.1	6.2	7.25	6.8
Fly	6.6	6.0	6.35	7.0	6.4
White	6.55	6.1	6.1	7.2	6.1
PMMA	5.5	3.6	3.9	6.4	3.9

tot. Ca^{2+}

time (months)	3	9.5	13.5	17	22	tot.
PC	63.8	266	105	117	160	712
BFS	149	340	120	119	149	877
Fly	61	225	120	101	250	757
White	52.8	275	131	132	255	845

tot.deposited (mg)

time (months)	3	9.5	13.5	17	22	tot.
PC	169	384	352	157	194	1256
BFS	112	380	281	225	231	1229
Fly	206	410.5	281	240	222	1359.5
White	220.5	389.5	253	265	144	1272
PMMA	385.5	790	594	357	407	2533.5

Amount of water collected in the container (l)

time (months)	3	9.5	13.5	17	22	tot.
PC	8.140	20.580	8.980	10.645	11.055	59.400
BFS	7.775	19.275	8.935	10.445	10.055	56.485
Fly	8.835	22.555	10.045	11.225	13.855	66.515
White	9.265	23.725	10.925	12.040	14.500	70.455
PMMA	12.705	30.280	13.945	14.805	17.210	88.945

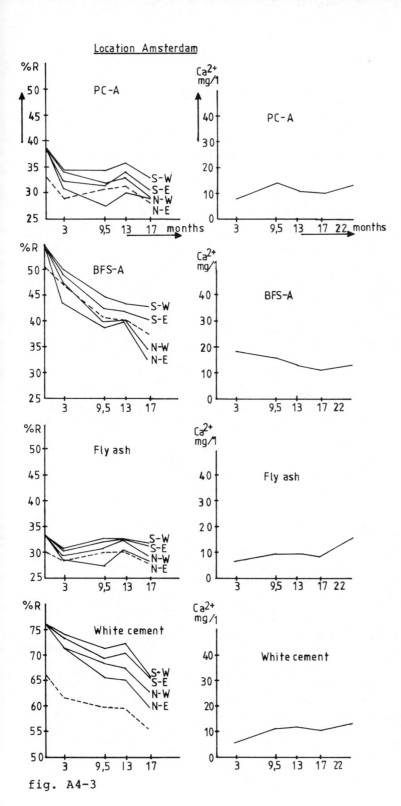

fig. A4-3

APPENDIX 5

Test methods on biological soiling

The test methods applicable in the case of soiling of
biological origin must offer a solution to two types of
problem, that is to say either determine the constituents
of this soiling or test the effectiveness of different
biocides.

Determination of constituents

The determination of the different constituents of soiling
can be undertaken by direct observation if the organisms
are sufficiently large in size, which is the case for
example of lichens. However, this method calls for tho-
rough knowledge of these organisms if precise determination is required, or reference can be made to books on
botany, such as the works of F.Dobson (Lichens, an illustrated guide, The Richmond Publishing Co. Ltd, 1981, 320p)
or P.Ozenda and G. Clauzade (Les Lichens, Etude Biologique
et Flore Illustrée, Masson et Cie, 1970, 801 p).
    When the soiling is due to microscopic organisms it is
impossible to identify them with the naked eye, except to
make shift with an overall impression (particularly from
colour). It is therefore necessary to apply methods used
in microbiology with cultures on specific or non-specific
media.
    As regards algae, an interesting method was proposed by
F. Grossin and P. Dupuy (Méthode simplifiée de déter-
mination des constituants des salissures : Colloque International UNESCO-RILEM Paris, 1978, 4.4. 41.p)

Test on biocides

It is difficult to determine in advance in the laboratory
the effect of biocides on soiling caused by different
types of microorganisms because it is not possible to
exactly reproduce the natural ageing conditions.
    Certain authors have proposed specific methods adapted
to the case of algae on stone and mortar (C. Grant, A.F.
Bravery Laboratory evaluation of algicidal biocides for
use on constructional materials; International Biodeterioration Bulletin 1981, 17, 4, pp 113-131) and the case of
fungi on paints (S. Barry, A.F. Bravery, L.J. Colman A
method for testing the mould resistance of paints; International Biodeterioration Bulletin 1977, 13, 2, pp 51-57).
    Such methods offer a comparative basis for determining
the effect of various biocides on the different soiling
components, but they remain dependent on laboratory conditions which are hardly representative of real exposure in

practice.

There is therefore a need to formulate more reliable testmethods based either on records of natural ageing in different sites or on accelerated ageing testing techniques.

APPENDIX 6

Outdoor exposure tests of atmospheric soiling

This investigation is concerned with the soiling rates of various building materials exposed on two sites at Delft.

The exposed materials are gypsum, concrete made with various types cement w.c. ratios, PVC, GUP, aluminium, stainless steel, window glass.

The atmospheric soiling of the exposed test specimens is caused by the accumulation of dust or other atmospheric substances deposited from the air directly on the material surface. On the façade of buildings the soiling will also be affected by substances which were first deposited on adjacent surfaces and then transported by rainwater to the surfaces concerned. This latter case of soiling by indirectly deposited particulate matter is not considered in this investigation.

The test specimens, consisting of metal and plastic sheets measuring 500 mm x 500 mm x 50 mm, are placed vertically in a rack and exposed facing south.

To observe the cleansing or staining effect of rain, the test specimens are placed in a sheltered and in an unsheltered rack on each site.

The two test sites are situated within a distance of a few hundred metres from each other, so that the climatic conditions are the same on both.

The arrangement of the exposed materials is also similar on both sites.

The difference between these two exposure tests lies in the difference between the heights of the test arrangements above ground level ; one site is at ground level and the other on the roof of a two-storey building.

Measurements

Elaborate atmospheric measurments on the site at ground level have been carried out by the Research Institute TNO, which will provide us with the required air quality data.

This exposure site has been selected because of these existing measurements.

The change in the appearance of the soiled material surface will be determined by reflectance measurements.

The measuring device has been developed by our measurement service group.

It consists of a light source (search unit), which illuminates the specimen surface, and a receptor containing a lens and photo-electric cell to measure the percentage of diffusely reflected light.

The reflectance value can be obtained by reading the meter on an instrument. A white and a grey card are used as a standard against which the reflection of the material

174

surface is measured. The cards used as standard reference
surfaces are obtained from Agfa, the white card is
assigned a reflectance value of 100 on the meter and the
grey card a value of 0.

The reflectance from the specimen surfaces will be read
between 100 and 0.

Calibration of the instrument will be done at the
start, and subsequently at sufficiently freqent intervals
during the operation, to ensure that the instrument
response is practically constant. To calibrate the instrument, it is adjusted to give the correct readings of the
standard cards.

After standarization, the sample is placed against the
search unit, and the reflectance value is read on the
meter.

Measurements are performed at nine points on the specimen surface so as to obtain an indication of uniformity.

Subsequent measurements are made initially, at
intervals of about 1 1/2 months, and of 3 months later on.

Degrees of soiling can also be determined by comparing
the test specimens with the reference specimens, which are
stored in a temperature and humidity controlled room in
the laboratory.

Preliminary test results

The tests at ground level started in February 1981 and the
tests above ground level in June 1981, so that only the
preliminary results of the tests at ground level are
available at present.

After an exposure period of 6 months it would appear at
first glance that the specimen surfaces have undergone no
significant changes due to deposition of air pollutants.

There are of course slight changes in the appearance of
the surfaces, but these are more due to the effects of
normal weathering processes.

Slight discolouring of some of the concrete or metal
samples can be observed ; some differences in the degree
of discolouring are also slightly visible.

The beginning of accumulation of dust is visible only
on the plastic sheets. Nor do the reflectance measurements
show significant differences between the reference reflectance value and the value after 6 months exposure. The
values after an exposure time of longer than 6 months are
given in the next tables.

Reflectance measurements (preliminary results)
1. Outdoor exposure at ground level
1.1. plastic and metal sheets

	PVC		GUP		Al.		stainless steel	
	wet	dry	wet	dry	wet	dry	wet	dry
Reference	20.0	20.0	75.0	75.0	28.1	28.1	39.1	39,1
7 months exposure	16.7	19.0	65.7	73.0	19.6	26.0	36.3	38.9
15 1/2 months exposure	18.0	10.0	69.3	72.0	20.1	24.9	35.8	30.9

1.2. Concrete slabs, made with portland cement, portland-slagcement and white cement.

	PC		PSC		WC	
	wet	dry	wet	dry	wet	dry
Reference	24.7	24.7	31.8	31.8	66.8	66.8
7 months exposure	13.7	22.1	36.3	32.6	54.6	66.7
15 1/2 months exposure	22.1	22.2	35.1	34.2	55.4	68.2

2. Outdoor exposure at roof level
2.1. Plastic and metal sheets

	PVC		GUP		Al.		stainless steel	
	wet	dry	wet	dry	wet	dry	wet	dry
Initial	20.0	20.0	75.0	75.0	48.9	49.2	31.4	32.2
7 1/2 months exposure	17.0	18.0	69.3	69.7	41.3	41.9	32.3	27.0
11 months exposure	19.0	19.0	70.7	72.3	41.2	44.8	32.4	29.2

	PC		PSC		WC	
	wet	dry	wet	dry	wet	dry
initial	37.6	37.8	30.9	29.6	65.4	65.9
7 1/2 months exposure	40.3	37.8	36.7	32.8	62.1	65.1
11 months exposure	43.2	39.0	40.3	33.6	66.1	67.7

Number of test specimens : one of each type
The values given in the table are the mean values of
measurements performed at 9 points on each specimen
surface.

APPENDIX 7

Water absorption test

To measure the water absorption of relative thick
materials such as concrete and glass reinforced concrete,
an absorptionmeter was developed by ir. L.G.W. Verhoef and
ir. B.F. van Leerdam from Delft University of Technology.
With the absorption meter (see fig. A7-1) it is possible
to gain an impression of the absorption of a material.
The method of measurement is non-destructive and can be
applied to all facades. The primary objective for the
development of the meter was to see if there is a relation
between the absorption capacity of the material of the
facade and its pollution. The relationship can be found
in the report "Texture, color and pollution" by ir. L.G.W.
Verhoef.

Procedure

- First is the mixing of the proper amount of polyether
  impression material (medium viscosity type 1) with a
  fixed amount of catalyst. The mixture solidifides
  fast, so that only a small amount needs to be made at
  once.
- The mixture is applied to the facade with the help of
  a ring. Because the ring is from polyether rubber,
  water cannot penetrate it but only the portion of the
  facade within the surface of the ring. The ring has
  a diameter of 100 mm contains a covered surface of
  $1/4$ 100 mm $= 7854$ mm .
- The edge of the absorptionmeter is sealed with an adhesive for a good closure between the absorptionmeter
  and the sealing ring.
- Place the absorptionmeter on the ring and seal it
  with polyether rubber. Careful work is necesarry to
  prevent leakage from the sealing ring which in turns
  determines the succes of the experiment.
- Bring the temperature of the water to the temperature
  of the facade and then the water is ready to be injected into the absorptionmeter.
- Fill the syringe with 5 ml of water. Be sure that no
  air bubbles remain in the syringe. The needle of the
  syringe is injected through the self-closing silicon
  rubber seal.
- The watersupply is opened so that the water can flow
  into the absorptionmeter. As soon as the water
  reaches the end of the capillair, the water supply is
  shut off. With the syringe, the bottom of the
  hollow meniscus is brought to the maximum height in
  the capillair.

hollow meniscus is brought to the maximum height in
the capillair.
- The measurement of the time begins with the starting
  of the stopwatch.
- When the material of the facade has absorbed so much
  water that the height in the capillair has declined
  to the minimum height in the capillair, the time is
  written down and the watersupply is brought back to
  the maximum height with the syringe.
- Continue   the last procedure, until 30 minutes have
  passed.

Other details

The absorptionmeter is made of glass so that the uneven
distribution of the water can be observed. The radius of
the capillair is 1.5 mm and was chosen because of the
following reasons:

- with a small diameter of the capillair, accurate
  readings of the amount of water that is absorbed can be
  made;
- the smaller the diameter of the capillair, the longer it
  takes for the water to reach the maximum height of the
  capillair. With a radius of 0.01 mm, it takes 10 hours
  for the water to reach the maximum height in the capillair; with a radius of 0.05 mm it takes one hour to
  reach the maximum height.
  The chosen radius of 1.5 mm allows the time for the
  maximum height to be reached in a time that is adjusta-
  ble for the experiment. The experiment is not influenced by the capillair rise time.
- if the diameter of the capillair is smaller, the rise in
  the capillair is higher.

The capillar rise h includes

$$h = \frac{2 . \gamma . \cos \phi}{r . s.} \qquad \text{with :}$$

= the surface tension of the liquid = 7.5 kg/mm ;
= the angle between the capillair wall and the surface
  of the solution
r = radius of the capillair ;
s = volume of the solution ; in this case water.

For water the formula is simplified to read :

$$h = \frac{2 . \gamma}{r} = \frac{15}{1.5} = 10 \text{ mm}$$

The maximum rise is 90 mm above the centre of the absorption meter. The minimum rise is 80 mm above the centre. The average height rise of the solution is 85 mm above the centre of the absorption meter. If the capillary rise of 10 mm is included, the experiment is carried out with a pressure of 75 mm.

The pressure in the column of 75 mm water can be seen in relation to the wind pressure that is acting on the facade. The Dutch regulations use $q = 1.2 \rho . v^2$ for wind pressure. v is the height, depending on the velocity of the wind measured during a one hour period with a 50% chance that this average is exceeded once in a five year period.

In the Netherlands, an average wind pressure value of $q = 800$ N/m^2 is assumed for building constructions with a height of 10-15 m. This wind pressure is equal to a water pressure height of 80 mm.

The water pressure of 75 mm is calculated from the centre of the absorption meter. The real pressure of the water varies between 25 and 125 mm. Because the absorption meter is a circular, the surface area around the centre is larger than the portion above or below (see fig.A7-2). As an example we have divided a circle into three parts with equal heights. The middle part occupies 41% of the surface of the circle compared to 29.5% for each of the other parts. This means that because of the hydrostatic pressure and the influence from the circle, the centre of gravity for the wall pressure is situated just below the centre of the absorption meter, whence the correlation with the windpressure is better.

Although there is a relation between the maximum windpressure at a certain height of the facade and the pressure on the facade in the middle of the absorption meter, this does not imply a relation with normal outside weather conditions. In general the actual wind pressure is lower than the maximum wind pressure. In addition, the hydrostatic pressure inside the meter is not uniform and therefore has a different influence on the rate of absorption.

A further problem has to do with the regulation of the moisture in the capillary system. Has the period before the measurement been dry or wet ? Is there an influence on the measurement from changes in relative humidity ?

A final problem concerns the length of time of the change in the capillary system due to the hydration and carbonation. The influence is made visible with the help of the absorption meter.

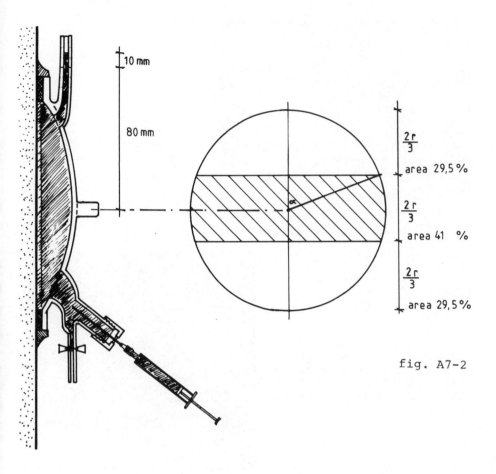

10 mm

80 mm

$\frac{2r}{3}$

area 29,5 %

$\frac{2r}{3}$

area 41 %

$\frac{2r}{3}$

area 29,5 %

fig. A7-2

fig. A7-1

photo 1
polyether rubber ring
applied to the facade

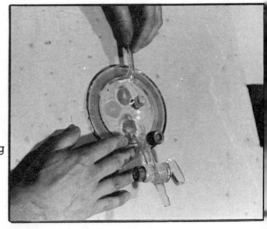

photo 2
the edge of the absorptionmeter is sealed with an adhesive. Absorptionmeter is
placed on the polyether ring

photo 3
absorptionmeter placed
on the polyether ring
and sealed with poly-
ether rubber. Careful
work is necessary to
avoid leakage

182

photo A
watersupply is opened
Water with the same
temperature as the fa-
cade is flowing in the
absorptionmeter. As the
water reaches the end of
of the capillair the
watersupply is shut off

photo 5
syringe with 5 ml of water
The needle of the syringe
is injected through the
self-closing silicon
rubber seal

photo 6
Starting the stopwatch
when the material of the
facade sucks so much
water that the minimum
height in the capillair
has reached, the time is
written down, and with
the help of the syringe
the water supply is
brought back to the max.
height.

Index

Abrasion resistance 155
Abrasive particles 131
Absorbancy 142
Absorption 104, 178
Absorption coefficient 138
Acids 128
Adhesive forces 93
Aerosols 65, 72, 76, 79, 81
Ageing 34, 52
Air 63, 65, 78, 83, 95, 112, 136, 162
Algae 111, 113, 114, 116, 118, 119, 172
Algicides 133
Alkaline ions 54
Alkalis 129
Aluminium 15, 41
Anisokinetic sampling 71, 73
Appearance 52
Asbestos cement 112
Atmosphere 18, 24, 63, 72, 76, 79, 81,
      83, 84, 85, 90, 94, 120, 174

Bacteria 111, 117, 119
Biocides 132, 172
Biological soiling 111, 119, 132, 172
Brick 8, 24, 50, 54, 55, 138, 144
Buildings 14, 34, 49, 140

Calcium hydroxide 54
Capillary forces 94
Carbon 15
Carbon dioxide 21
Cellulose 27
Cement 138, 162, 165
Chalking 53, 55
Chelates 130
Chemical changes 26
Chemical cleaning 125, 128

Chemical compounds 130
Chimneys 85
Chlorides 21
Cleaning 124
Climate 137
Climatic ageing 34
Coarse particles 69, 72
Cohesive forces 93
Components 34, 109
Composition 136
Concrete 2, 21, 49, 54, 56, 106, 138,
      139, 154, 157, 162
Condensation 95
Coning 90
Conservation 124
Copper 15
Corrosion 24, 30, 38, 40, 44, 46
Cracking 47
Cysts 113

Decohesion 53, 55
Degradation 27, 30
Deposition 66, 77
Design 58, 136
Detailing 61, 140, 147
Development 115, 138
Diffusion 86, 88, 91
Dirt 62, 127
Driving rain 99, 104, 107, 138
Dry blasting 131
Dry deposition 19, 66
Dust 94, 95

Efflorescence 49, 50, 54
Electric forces 94
Exposure 174

Façades 1, 49, 72, 83, 92, 124, 136, 138, 142, 144
Fanning 90
Filiform 44
Film 15
Flashing 102
Fungi 111, 113, 114, 116, 119
Fungicides 133

Galvanic corrosion 42
Gases 77
Glass 12, 30, 51, 54, 56, 144
Gravity 93
Grinding 106
Gutters 149

Hardness 16
High Volume Sampling 71
Hydraulic models 89

Iron 15, 50

Javel water 132

Lichens 112, 114, 116, 119
Light 27, 116
Light-absorption 67, 68, 76
Location 136
Lofting 91
Looping 9

Materials 138
Measurement 70, 73, 165, 174, 176
Mechanical cleaning 125, 131
Mechanical properties 52
Metal ions 54
Metals 15, 37, 55, 56
Methods 127
Microorganisms 111, 112, 114, 115, 117, 119, 132
Mineral soiling 120
Modelling 88
Molecular forces 93
Moss 112

Numerical models 88
Nutritive conditions 116

Old buildings 59
Outdoor 174

Paints 53, 55, 56, 119, 133, 138
Particle size 66, 79, 80, 81
Particles 76, 77, 92, 112, 126
Particulates 65
pH values 115
Physico-chemical properties 52
Pigments 55
Pitting 41, 46
Plants 111
Plastics 14, 34, 52, 55, 56, 112, 119, 138
Pollutants 69, 70, 72, 75, 85, 86, 88, 136
Pollution 24, 63, 65, 83
Polymethylacrylate 162, 165
Porosity 16
Pressure 180
Primary soiling 72
Profiling 140
Protection 40
Putty 154
Pyrites 50

Rain 94
Raindrops 98, 100, 104
Rainfall 55, 98, 104, 106
Rainwater 19, 143
Reactive gases 19
Reflectance 165, 176
Rendering 138
Repainting 59
Restoration 124
Reversal 91
Ribs 145
Roofing materials 112
Roughness 16
Run-off 55, 98, 104, 106, 119, 138
Rust 49

Salts 50
Sampling 71, 73, 155, 157
Sandblast 155, 157
Scavenging 77
Sealants 56
Secondary soiling 72

186

Semantic differential 62
Silicates 54
Simulation 88
Skin 1, 2, 8, 10, 12, 14, 15, 140
Slates 112
Smoke 91
Soiling 35, 58, 70, 72, 109, 110, 111,
   119, 124, 132, 162, 172, 174
Solar radiation 116
Solarization 30
Soot 67, 68, 72, 81
Splashes 102
Spores 112, 113, 133
Stainless steel 15, 46
Stains 49, 50
Steam cleaning 127
Steel 15, 47, 109
Stone 8, 24, 50, 54, 56, 138, 139
Stress 16, 47
String course 141
Strip 141
Substrates 92, 117, 132
Sulphate-reducing bacteria 117
Sulphooxidising bacteria 117
Sulphur dioxide 21
Surface 15

Surfaces 95, 104

Temperature 83, 84, 91, 116
Tests 155, 157, 165, 172, 174, 178
Total Suspended Particulates 71
Turbulence 95

Ultraviolet light 26
Urban concentrations 75

Vanadium 50
Vapour 21

Walls 104
Washing 125, 127, 143
Water 21, 84, 94, 115, 119, 127, 131,
   133, 178, 180
Weather 83
Weathering 21, 24, 25, 30, 34, 37, 47,
   55, 58
Wet blasting 131
Wet cleaning 127
Wet deposition 19, 66
Wind 83, 86, 88, 94, 95, 120, 137, 180
Windows 142
Wood 10, 25